bibliographia geodaetica

AKADEMIE-VERLAG · BERLIN

Volume 22 · 1984 · Number 3

ISSN 0006-1239 bibl. geod., Berlin 22 (1984) 3, 189-292 EVP 15,- M

CONTENTS

BIBLIOGRAPHIA GEODAETICA

International Geodetic Documentation

Booklet of reports of the
International Association of Geodesy
International Federation of Surveyors
Technical University of Dresden, GDR

Published by the National Committee of Geodesy and Geophysics of the Academy of Sciences of the German Democratic Republic

Prof. Dr.-Ing. habil. Lothar Stange, president

Edition by the Centre of International Documentation of Geodesy, Technical University of Dresden, Section Geodesy and Cartography

Dr.-Ing. Heinz Finger, manager

AKADEMIE-VERLAG · BERLIN

bibliographia geodaetica	Volume 22 1984 No. 3	p. 189—292	Berlin 1984

Sources of supplies

Orders and subscriptions of the "bibliographia geodaetica" should be sent:

- in the German Democratic Republic to a book-shop, or to the Akademie-Verlag, DDR – 1086 Berlin, Leipziger Str. 3–4;
- in socialist countries to a book-shop for foreign-language literature, or to the responsible post-office;
- in the Federal Republic of Germany and Berlin (West) to a book-shop, or to the supply depot KUNST UND WISSEN, Erich Bieber OHG, Wilhelmstr. 4–6, D – 7000 Stuttgart 1;
- in other West-European countries to a book-shop, or to the supply depot KUNST UND WISSEN, Erich Bieber GmbH, Dufourstr. 51, CH – 8008 Zürich;
- in the remaining foreign countries to the International Book and Journal Trade; to the Buchexport, Volkseigener Aussenhandelsbetrieb der Deutschen Demokratischen Republik, DDR – 7010 Leipzig, Postfach 160, or to the Akademie-Verlag, DDR – 1086 Berlin, Leipziger Str. 3–4.

Information of additional sources of supplies is given by the Bureau Central de l'AIG, 39ter, rue Gay Lussac, F – 75005 Paris.

bibliographia geodaetica

Herausgeber: Nationalkomitee für Geodäsie und Geophysik der Akademie der Wissenschaften der Deutschen Demokratischen Republik.
Verlag: Akademie-Verlag, DDR – 1086 Berlin, Leipziger Straße 3–4; Fernruf: 2 23 62 21 oder 2 23 62 29; Telex Nr.: 11 44 20; Bank: Staatsbank der DDR, Berlin, Kto.-Nr.: 6836-26-20712.
Chefredakteur: Dr.-Ing. Heinz Finger.
Redaktion: Zentralstelle für Internationale Dokumentation der Geodäsie, Technische Universität Dresden, Sektion Geodäsie und Kartographie, DDR – 8027 Dresden, Mommsenstraße 13; Fernruf: 4 63 28 23 (RIS 1107).
Druckgenehmigung: Ag 451/48/84.
Satz und Druck: Druckwerkstätten Stollberg VOB, DDR – 9150 Stollberg (Erzgeb.) - III/6/50 684 (415).
Erscheinungsweise: Die „bibliographia geodaetica" erscheint jeden zweiten Monat. Bezugspreis eines Jahrganges 120,– M zuzüglich Versandspesen (Preis für die DDR 90,– M).
Bestellnummer dieses Heftes: 1073/22/3.

bibliographia geodaetica

New series, Volume 22, No. 3 1984

AIDOS Systematics **12.01** **GENERALITIES**

General problems of geodesy, surveying, photogrammetry and cartography

Management, steering, cooperation with other branches of economy
History of development
National scientific-technical organizations
International co-operation
Research, invention affairs, innovators
Information activities, reference books, terminology
Standardization, metrological ensuring
Organization, planning, economy
Staff and training
Quality proof and quality ensuring
Labour and health safety

ALLGEMEINES

Allgemeine Fragen der Geodäsie, Photogrammetrie und Kartographie

Leitung, Zusammenarbeit mit anderen Wirtschaftszweigen
Historische Entwicklung
Nationale wissenschaftlich-technische Organisationen
Internationale Zusammenarbeit
Forschung, Erfindungswesen, Neuerer
Informationstätigkeit, Nachschlagewerke, Terminologie
Standardisierung, metrologische Sicherstellung
Organisation, Planung, Ökonomie
Kader und Ausbildung
Qualitätskontrolle, -sicherung
Arbeits- und Gesundheitsschutz

528.236(091)(439)

Bartha, L. jr.
A legkorabbi hazai foeldrajzi koordinatakrol
Geod. es Kartogr. — Budapest 35 (1983) 5. — p. 353—359 : 2 tab. 25 lit.
On Hungary's oldest geographical co-ordinates (HUN)
AS 12.01.09
Key words: Astronomical observation, co-ordinate table, geographical latitude, geographical longitude, historical co-ordinates, renaissance

The "Geography" of Ptolemy shows the geographical latitude and longitude of about 70 places from the area of the Carpathian Basin. From these the elevation of the Pole at Scarabantia (Sopron) and Sirmium (Mitrovica, Yugoslavia) has been based on a real astronomical observation, i.e. on the determination of the longest summer day. From the times of the early renaissance the co-ordinate table of magister Reinhardus in Klosterneuburg (1422—1436) was the first one which showed the positions of a relatively great number of places in Hungary. The author listed tables showing their finding place and their library reference and containing data of Hungarian towns. He describes 7 more important tables up to the end of the fifteenth century.

Veroe-Hetenyi, National Correspondent, Hungary **IGD 84/A 39**

378.6(47+57):528(091)

Bol'shakov, V. D.
Vysshee geodezicheskoe obrazovanie v SSSR za 60 let i rol' MIIGAiK v ego razvitii
Geod. i Kartogr. – Moskva (1982) 12. – p. 29–34
The higher geodetic education in the USSR for 60 years and the role of the Moscow Institute of Engineers of Geodesy, Aerial Survey and Cartography (MIIGAiK) in its development (RUS)
AS 12.01.79
Key words: Geodetic education, USSR

The geodesists in the USSR are trained in 22 higher educational establishments (two specialized in Moscow and Novosibirsk, geodetic departments and sections of other educational establishments) in specialities as follows: astro-geodesy, marine geodesy, applied geodesy, space geodesy, aerophotogeodesy, map publication, design and compilation of maps, reproduction technique, optical and optoelectronical instrument construction, study of natural resources. Every year up to 700 specialists graduate from the MIIGAiK. Among the specialists graduated from MIIGAiK there are foreigners from 40 countries. In MIIGAiK 125 doctor's and 750 master's degrees have been awarded. Since 1976 a faculty for improving the qualification of teachers in geodetic disciplines exists.

Dobrokhotov, National Correspondent, USSR IGD 84/A 40

061.3(100)FIG"1983"(497.2):528

Draheim, H.
XVII. Kongress der Federation Internationale des Geometres (FIG)
Allg. Vermess.-Nachr. – Karlsruhe 90 (1983) 11–12. – p. 417–427
XVII Congress of the Federation Internationale des Geometres (FIG) (GER)
AS 12.01.17/12.23.05/12.23.09
Key words: FIG Congress 1983

Techn. Univ. Dresden IGD 84/A 41

061.3(100)IUGG"1983"(430-2.2):528

Draheim, H.
XVIII. Generalversammlung der Internationalen Union fuer Geodaesie und Geophysik (IUGG)
Allg. Vermess.-Nachr. – Karlsruhe 90 (1983) 11–12. – p. 427–429
XVIII General Assembly of the International Union of Geodesy and Geophysics (IUGG) (GER)
AS 12.01.17/12.15.09
Key words: IUGG General Assembly XVIII

Techn. Univ. Dresden IGD 84/A 42

528.2:629.78

Hieber, S.
European space agency – ESA (ENG)
CSTG Bull. – (1982) 4. – p. 131–139
AS 12.01.17/12.15.17
Key words: Cosmical geodesy, Europe, remote sensing, scientific-technological societies of non-socialist states, trend of development

Accounts of projects in the field of cosmical geodesy and geodynamics planned by the ESA till 1995.

Kombinat Geod./Kart. IWT, Leipzig IGD 84/A 43

061.3(100+437)"1982":523/525

Jurkina, M. I.
Mezhdunarodnyjj simpozium «Figury Zemli, Luny i drugikh planet»
Geod. i Kartogr. — Moskva (1983) 3. — p. 54—55
The international symposium "The figures of Earth, Moon, and other planets" (RUS)
AS 12.01.17/12.15.09/12.15.13
Key words: Celestial bodies, geodetic astronomy, international symposium, physical geodesy

The symposium (Prague, 20—25 September, 1982) got together the participants from CSSR, USSR, Hungary, GDR, Bulgary, Poland, Egypt, Denmark, Italy, USA, FRG, Sweden, and Yugoslavia. The topics of the papers included such themes as:

1) the force function of the system of celestial bodies, dynamics of the system "Earth-Moon", dynamics of deformable Earth
2) the boundary value problems of physical geodesy, the methods of their solution
3) representation of gravitational fields of celestial bodies
4) comparative studies of the figures (gravitational fields) of the bodies of the Solar system.

It is pointed out that the available dynamic theories of translational-rotational motion of celestial bodies are not sufficient for explanation of the data obtained.

Dobrokhotov, National Correspondent, USSR IGD 84/A 44

528.38(091)

Meshherskijj, I. N.
Glavnaja vysotnaja osnova strany
Geod. i Kartogr. — Moskva (1983) 1. — p. 22—26 : 6 lit.
The principal height datum of the country (RUS)
AS 12.01.09/12.01.37
Key words: Height datum

The main stages of establishing the first- and second-order levelling nets of the USSR are discussed. While in the period from 1871 to 1917 45.5 thousands of km of levelling lines were established, so in the period from 1923 to 1943 there were over 114,000 km.

The adjustments of the levelling net were carried out in 1894, 1933, 1950 and 1977. The R.M.S. errors of height determination with respect to the Kronstadt water-gauge as obtained from 1977 adjustment are as follow: Novosibirsk 62 mm, Yakutsk 93 mm, Vladivostok 112 mm, Pevek 142 mm. Therefore, the first- and second-order levelling net allow reliable determinations of heights in a single system over the whole territory of our country. The detailed accuracy parameters of the first-order net established after 1945 are given.

Dobrokhotov, National Correspondent, USSR IGD 84/A 45

528:502.7

Novozhenin, Ju. Kh.
Geodezija na sluzhbe okhrany prirody
Geod. kontrol' v str-ve. — Kujjbyshev (1982). — p. 88—90 : 2 lit.
Geodesy at the service of nature protection (RUS)
AS 12.01.05
Key words: Environment, geodesy, protection of nature

Dobrokhotov, National Correspondent, USSR IGD 84/A 46

061.3(100+497.2)"1983"FIG:528.4/.5

. . .

Berichte ueber den XVII. Internationalen Kongress der Vermessungsingenieure vom 19.—28. Juni 1983 in Sofia, Bulgarien
Vermess. Photogramm. Kulturtechnik. — Zuerich 81 (1983) 10. — p. 370—379
Reports on the XVII International Congress of Surveying Engineers from June 19 to 28, 1983 in Sofia, Bulgaria (GER)
AS 12.01.17
Key words: FIG, International Congress, Sofia

Commission 1: Vocational practice, organisatory and legal techniques
Commission 2: Vocational training and technical literature
Commission 3: Land information systems (LSI)
Commission 4: Hydrographic surveyings
Commission 5: Surveying instruments and methods
Commission 6: Engineering surveying
Commission 7: Land registry and land improvement
Commission 8: Urban and regional planning and its realization
Commission 9: Valuation of real property and property market

Techn. Univ. Dresden IGD 84/A 47

061.24(100):528.27

. . .

The 11th Meeting of I G C Hamburg 10—13 Aug. 1983 (ENG/FRE)
Bull. Inform. — Toulouse (1983) 53. — p. 27—217
AS 12.01.17/12.15.09
Key words: IAG General Assembly, Hamburg 1983, Meeting of International Gravity Commission

Contents:

- Program
- List of Participants
- II.1. Minutes of WG. 1, WG. 2, WG. 3, WG. 4 and D.B. and Activity Report 1979—1982
- II.2. Reports and Papers Presented at I.G.C.

Techn. Univ. Dresden IGD 84/A 48

061.3(100)"1982"(437.11):(523.3/.4+525+528)
061.12(430.2+437+438+439+47+497+498+57)CAPG:528

. . .

Proceedings of the International Symposium "Figure of the Earth, the Moon and other planets" (ENG)
Commission of the Academies of Sciences of the Socialist Countries for Multilateral Co-operation on the Complex Problem "Planetary Geophysical Investigation" et al., Prague, Czechoslovakia from September 20 to September 25, 1982 (Ed. P. Holota)
Publ. Res. Inst. Geod., Topogr. and Cartogr. Prague. — Prague (1983). — 493 p. : div. fig. div. tab. div. lit.
AS 12.01.17/12.15.09/12.15.13
Key words: CAPG, figure of the Earth, geodetic astronomy, moon, physical geodesy, planets, symposium

Contents:

1. Force function of the system of celestial bodies
   Dynamics of the Earth-Moon system
   Dynamics of the deformable Earth (p. 1—114)

2. Boundary value problems in physical geodesy
   Further development of methods of their solution (p. 115—286)
3. Representation of gravity fields of celestial bodies (p. 287—430)
4. Comparative studies of figures (gravity fields) of solar system bodies (p. 431—486)

Techn. Univ. Dresden IGD 84/A 49

061.3(100+480)"1982"ISPRS/Com. III:528.7/.8

. . .
Proceedings of the Symposium "Mathematical models, accuracy aspects and quality control" (ENG)
Internat. Soc. Photogramm. and Remote Sens., Commiss. III,
June 7—11, 1982
Helsinki Univ. of Technology, Otaniemi, Finland. — Otaniemi (1982) 24-III IAP. — 640 p. : div. fig. div. tab. div. lit.
(Ed. by Hakkarainen, J.; Kilpelae, E.; Savolainen, A.)
AS 12.01.17/12.29.21/12.29.25
Key words: Digital terrain model, gross error, image analysis, photogrammetric triangulation, symposium, systematic error

Commission III "Mathematical analysis of data" of ISPRS with the Working Groups

1: Identification and elimination of gross and systematic errors

2: On-line photogrammetric triangulation

3: Mathematical aspects of digital terrain information

4: Mathematical aspects of image registration, rectification and enhancement

5: Mathematical pattern recognition and image analysis

realized at Finland, June 7—11, 1982 in a symposium.

Techn. Univ. Dresden IGD 84/A 50

061.3(100+489)"1982"FIG:528.3
528.33.088:528.14:681.3

. . .
Proceedings "Survey control networks" (ENG)
Internat. Federation of Surveyors — FIG —, Meeting of Study Group 5B, 7th—9th July, 1982, Aalborg Univ. Centre, Denmark
(Ed. K. Borre; W. M. Welsch)
Schr.-R. Wiss. Studieng. Vermess.-Wesen Hochsch. Bundeswehr Muenchen. — Muenchen (1982) 7. — 428 p. : div. fig. div. tab. div. lit.
AS 12.01.17/12.15.25/12.23.05
Key words: Adjustment, geodetic network, symposium

The Study Group 5B of International Federation of Surveyor (FIG) realized at Aalborg, Denmark from 7th—9th July, 1982 a meeting with five main topics:

- Data processing and adjustment
- Interaction of stochastic and functional models
- Quality of networks
- Network densification
- Optimization of control networks.

Techn. Univ. Dresden IGD 84/A 51

061.3(430.1)"1983":528

. . .
„Vermessungswesen und Lebensraum – Erfassen, Darstellen, Gestalten"
Vortraege und Berichte vom 67. Deutschen Geodaetentag Hannover 1983
Z. Vermess.-Wesen. – Stuttgart 108 (1983) 12. – p. 521–600 : div. fig. div. tab.
"Surveying and living space – registration, representation, design" (GER)
AS 12.01.13
Key words: Meeting of the geodesists of the FRG 1983

Techn. Univ. Dresden IGD 84/A 52

Technical University of Dresden
Section Geodesy and Cartography

bibliographia geodaetica

AIDOS Systematics **12.15** **GROUP I GEODESY**

Geodesy, geodetic measuring techniques and their treatment

GRUPPE I HÖHERE GEODÄSIE

Höhere Geodäsie, geodätische Meßverfahren und ihre Bearbeitung

AIDOS Systematics **12.15.05** **Group I Geodesy**

Mathematical geodesy

Co-ordinate determination, co-ordinate transformation, reference bodies and systems used for computing processes
The Earth as a sphere, a spheroid, an ellipsoid . . .
Mapping of the Earth body on reference surfaces
Relations between the figure of the Earth and reference surfaces
Solving of geodetic main problems

Gruppe I Höhere Geodäsie

Mathematische Geodäsie

Koordinatenbestimmung, -transformation
Für Rechenprozesse verwendete Körper und Systeme
Erde als Kugel, Sphäroid, Ellipsoid . . .
Abbildung des Erdkörpers auf Näherungsflächen
Beziehungen zwischen Erdfigur und Näherungsflächen
Lösung geodätischer Hauptaufgaben

528.235.1(620)

Abdelhamid Kamal Hassan / Szadeczky-Kardoss, Gyula
Simple determination of the convergence of the meridians for the area of Egypt (ENG)
Acta Geodaet., Geophys. et Montan. – Budapest 18 (1983) 1–2. – p. 59–65 : 3 fig. 1 tab. 5 lit.
AS 12.15.05
Key words: Convergence of meridian, co-ordinate system, Egypt, Gauss-Krueger co-ordinates, Hayford ellipsoid, meridian arc distance, North Africa, universal transversal mercator (UTM)

The paper presents a simple computation method for the determination of the convergence of the meridian for the area of Egypt, i.e. for the territory enclosed between the geographical latitudes of 22°–32° north of the equator. The method is valid for the Hayford 1909 ellipsoid, and the equations are suitable to be used for the universal transversal mercator (UTM) projection system. A numerical example is given to illustrate the practical use of the attached table. The present formula gives the convergence of meridian with an accuracy of 0.1 second of arc which is suitable for geodetic computation tasks. The table of coefficients facilitates the execution of practical computations. With these, the computation work is considerably shortened and made suitable for rapid use to complete the up-to-date gyrotheodolite measurement for orientation purposes.

Veroe-Hetenyi, National Correspondent, Hungary IGD 84/I 288

528.223:517.94

Arnold, Kurt
Zur Differentialgleichung der physikalischen Geodaesie
Gerlands Beitr. z. Geophysik. — Leipzig 92 (1983) 6. — p. 449—464 : 1 fig. 15 lit.
Differential equation of physical geodesy (GER)
AS 12.15.05/12.15.09/12.15.37
Key words: Differential equation, physical geodesy

The linear shape of the differential equation of physical geodesy is considered. The reference potential is generalized, it includes not only the spherical harmonics of second order but also the higher order spherical harmonics. The uniqueness of the solution is proved for a star-shaped Earth. If the real gravity potential fulfils a precisely formulated condition, the uniqueness is shown to be free from a reference potential.
(Author's summ.)
Techn. Univ. Dresden IGD 84/I 289

517.927:528.2

Arnold, Kurt
The mixed boundary value problem of geodesy (ENG)
Gerlands Beitr. Geophysik. — Leipzig 92 (1983) 5. — p. 391—406 : 1 fig. 16 lit.
AS 12.15.05/12.15.37
Key words: Boundary value problem

The matrix for the solution of Stokes's boundary value problem is fitted to the solution of the mixed boundary value problem. The uniqueness of this solution is proved. This question is extended to the Molodensky boundary value problem of mixed type. It is shown that the solution of this problem obviously is likewise unique.
(Author's summ.)
Techn. Univ. Dresden IGD 84/I 290

528.236

Csepregi, Sz. / Soha, G.
Szabatos vetueleti szamitasok
Geod. es Kartogr. — Budapest 35 (1983) 4. — p. 247—257 : 7 fig. 1 tab. 13 lit.
Precise projection calculations (HUN)
AS 12.15.05
Key words: Arc-to-chord reduction, convergence of meridians, cylindric projection, hyperbolic function, stereographic projection

Motivated by modern computing techniques, the authors examined the solutions and problems of projections. Their aim was to use closed formulae and direct plane co-ordinates for arc-to-chord reduction, the scale factor and the convergence of meridians, furthermore that of the conversion between neighbouring zones of projection. Authors report collectively on their results of analyses concerning stereographic and cylindric projections. These results show the important role of the formula of spherical excess, the suitable employment of the system of geographic co-ordinates as an aid, the use of hyperbolic functions and a transformation accomplished by geometric equivalence.
Veroe-Hetenyi, National Correspondent, Hungary IGD 84/I 291

528.232.24

Ehlert, Dieter
Die Bessel-Helmertsche Loesung der beiden geodaetischen Hauptaufgaben
Z. Vermess.-Wesen. — Stuttgart 108 (1983) 11. — p. 471—478 : 2 fig. 7 lit.
Bessel-Helmert solution of the principal problems of geodesy (GER)
AS 12.15.05/12.33.13
Key words: Coordinates transformation

By definition, "Umbezifferung" means a kind of transformation, changing the grid from one map projection into another by improving the distortion. Up to now investigations have been mainly related to area distortion. In this paper angular distortions are also considered and the conditions for conformal projections are given, because of the particular interest they find today in property surveying and land information systems. In this connection not only the transformation of the grid, but also that of the survey points is of interest, mainly in view of analytical and numerical methods and data processing. Finally it is shown how the transformation method, called "Umbezifferung", can be interpreted as projection directly from the original, explained by an example.

(Author's summ.)

Techn. Univ. Dresden IGD 84/I 292

528.21:528.236

Fys, M. M.

O vychislenii model'nykh stokosvykh postojannykh Zemli, sootvetstvujushhikh predstavleniju o ejo plotnosti, chastichnojj summojj obobshhjonnogo rjada Fur'e

Geod., Kartogr. i Aehrofotos"jomka. — L'vov (1982) 36. — p. 103—107 : 4 lit.

On computation of the model Stokes' constants of the Earth that correspond to its density representation by a partial sum of the generalized Fourier series (RUS)

AS 12.15.05/12.15.09

Key words: Mathematical geodesy, Stokes' constant

The formulas for computation of Stokes' constants of a three-axial ellipsoid from the coefficients of expansion of its mass density into a series using a system of polynomials are developed.

Dobrokhotov, National Correspondent, USSR IGD 84/I 293

528.236

Ganeko, Yasuhiro

Transformation of the coefficients of spherical harmonics expansion (JAP)

J. Geod. Soc. Japan. — Tokyo 29 (1983) 3. — p. 166—167 : 2 fig. 2 lit.

AS 12.15.05/12.15.09

Key words: Mathematical geodesy, physical geodesy

The expansion of geophysical data distributed on the Earth's surface is generally given by spherical harmonics series in which the z-axis of the polar co-ordinate is taken to be the rotational axis of the Earth. For some regional mathematical treatment of such geophysical data, the expansion into spherical harmonics series whose pole is located at an arbitrary point is sometimes very convenient. The transformation equations of the coefficients of spherical harmonics series are given here.

Nagasawa, National Correspondent, Japan IGD 84/I 294

528.235.1

Hazay, Istvan

Quick zone-to-zone transformation in the Gauss-Krueger projection (ENG)

Acta Geodaet., Geophys. et Montan. — Budapest 18 (1983) 1—2. — p. 71—81 : 1 fig. 3 tab. 10 lit.

AS 12.15.05

Key words: Arc-to-chord reduction, conform transversal cylinder, ellipsoidic computations, Gauss-Krueger co-ordinates, projections, scale factor, universal transversal mercator co-ordinates

The Gauss-Krueger projection is used in many countries and there are several methods for zone-to-zone transformations for its adjacent zones. These methods are generally rather tedious. The author found recently an immediate connection which radically simplifies the problem. Neither arc-to-chord reductions, nor scale factors are necessary as the difference between the data of the two zones can be easily determined. The connection is based on the two values of the y co-ordinates at the border meridian.

The method can also be used for the connecting systems of any oblique mercator (cylindrical) projection of the sphere, and the method is in this case even simpler due to the less complicated curvature of the sphere, here the connection depends on the distance of the two y axes.

Veroe-Hetenyi, National Correspondent, Hungary IGD 84/I 295

528.232.22

Laping, K. A.
Vychislenija dliny dugi normal'nogo sechenija mezhdu antipodnymi tochkami
Geod. i Kartogr. — Moskva (1983) 3. — p. 13—15 : 3 lit.
Computation of the length of a normal section arc between antipodal points (RUS)
AS 12.15.05/12.15.37
Key words: Geodesic

A method of computation of the said length with accuracy of 1 mm is suggested. The method provides for procedures as follow: computation of geodetic co-ordinates of intermediate points of the arc with spacing of 0.2° (900 points in total); computation of normal section arcs between adjacent intermediate points by assuming them to be circular ones; calculation of the total normal section arc by summing up the lengths ot these individual arcs. The solution of one variant of the problem requires approximately 10 s of computer time.

Dobrokhotov, National Correspondent, USSR IGD 84/I 296

528.23:629.783

Petroshkjavichjus, P. A. / Putrimas, R. S. / Ramanauskas, R. L.
O skhodimosti rjadov nevozmushhjonnogo blizparabolicheskogo dvizhenija
Tr. po geod. Vil'njus. inzh.-stroit. in-t. — Vil'njus (1982) 11. — p. 22—24 : 3 lit.
On convergence of series of undisturbed quasiparabolic motion (RUS)
AS 12.15.05/12.15.17
Key words: Mathematical geodesy, satellite geodesy

Dobrokhotov, National Correspondent, USSR IGD 84/I 297

528.235.1

Varga, J.
Conversions between geographical and transverse mercator (UTM, Gauss-Krueger) grid co-ordinates (ENG)
Period. Polytechn. : Ser. Civil Engin. — Budapest 27 (1983) 3—4. — p. 239—251 : 1 fig. 1 tab. 12 lit.
AS 12.15.05/12.15.37
Key words: Co-ordinate conversion, geographical co-ordinates, grid co-ordinates

Most of difficulties in mutual computer conversions between universal transverse mercator (UTM) grid and geographical co-ordinates arise in determining S (the meridian arc length from the equator) and Φ_1 (the ellipsoidal geographical latitude of the foot point which is on the central meridian).

Effective methods for such conversions are suggested, suiting pocket calculators with less storage capacities, just as are programs for calculators type HP 67/97 suitable for conversions on the entire surface of any ellipsoid, as indicated in the title.

Veroe-Hetenyi, National Correspondent, Hungary IGD 84/I 298

528.235.1

Varga, Jozsef
A Lambert-fele szoegtarto kupvetueletroel
Geodez. es Kartogr. — Budapest 35 (1983) 1. — p. 25—30 : 7 fig. 7 lit.
The Lambert conformal conic projection (HUN)
AS 12.15.05/12.15.37
Key words: Conformal conic projection, convergence, co-ordinate transformation, Lambert projection

The author shows unified equations for the case of one standard parallel and of two standard parallels. Program is written for use on Hewlett-Packard 67/97 calculators computing on any spheroid from geographic co-ordinates to plane co-ordinates, convergence and scale factor and also from plane co-ordinates to geographic co-ordinates, convergence and scale factor.

Veroe-Hetenyi, National Correspondent, Hungary IGD 84/I 299

528.236

Vasil'ev, A. B. / Shupta, N. S.
O nekotornykh svojjstvakh koordinatnykh preobrazovanijj, primenjaemykh v geodezicheskikh vychislenijakh
Inzh.-geod. raboty pri str-ve, montazhe i ehkspluat. inzh. sooruzh. — Moskva (1982). — p. 145—156 : 8 lit.
On some properties of co-ordinate transformations used in geodetic computations (RUS)
AS 12.15.05
Key words: Co-ordinates transformation, mathematical geodesy

For the optimal approximation of a physical space of geodetic measurements by vector spaces the differences in accuracy of the initial parameters are to be taken into account. When increasing the degree of the approximating polynomial the variations of all previously computed coefficients of approximation including the translation of the origin in transformation of rectangular co-ordinate systems should be accounted for, thereby the errors of determination of translation elements increases with the rotation angle of the systems to be transformed.

Dobrokhotov, National Correspondent, USSR IGD 84/I 300

528.232

Wolfrum, O.
Die Reduktion einer Raumstrecke auf den ellipsoidischen Normalschnittbogen
Allg. Vermess.-Nachr. — Karlsruhe 91 (1984) 1. — p. 23—33 : 2 fig. 10 lit.
The reduction of a spatial distance to the normal section curve of the ellipsoid (GER)
AS 12.15.05
Key words: Ellipsoidal normal section, mathematical geodesy

A Fourier series is developed which gives the length of the normal section curve as a function of the angle of depression which in turn by applying a simple iteration procedure is found from the chord length. Thus, slant ranges of any lengths can be reduced to the ellipsoid surface with an error less than one millimetre. In due course, a solution for the reverse case of geodetic co-ordinate computation is given.

(Author's summ.)

Techn. Univ. Dresden IGD 84/I 301

061.3(100)+45"1978":528.23

. . .
VII Simposio sulla Geodesia Matematica; Assisi, 8–10 Giugno 1978
Firenze, 1981. – Commiss. Geod. Italiana, Gruppo Nazionale Geofis., Terra Solida del Consiglio Nazionale Ricerche. – 308 p.
Seventh Symposium on Mathematical Geodesy (4th Hotine Symposium); Proceedings; Assisi, Juni 8th to Juni 10th, 1978 (ENG)
AS 12.15.05/12.01.17

Key words: 4th Hotine symposium, mathematical geodesy, proceedings

Contents: Part 1

Grafarend, E. W. – Criterion matrices (p. XI)

Grafarend, E. W. – Anholonomity of gravity space co-ordinates (p. XI)

Blaha, G. – The number of parameters in a geopotential model containing spherical harmonics and point masses (p. XII)

Marussi, A. – The tidal field of a planet and the related intrinsic reference systems (p. XIII)

Chovitz, B. H. – Perspective projections in terms of the metric tensor to the second order (p. XIV)

Grafarend, E. W. et al. – Concept for reference frames in geodesy and geodynamics: the reference directions (p. XV)

Rummel, R.; Gerstl, M.; Reinhart, E. – Stability control and the use of bounds for downward continuation problems (p. XVI)

Techn. Univ. Dresden IGD 84/I 302

061.3(100)+45"1978":528.23

. . .
VII Simposio sulla Geodesia Matematica; Assisi, 8–10 Giugno 1978
Firenze, 1981. – Commiss. Geod. Italiana, Gruppo Nazionale Geofis., Terra Solida del Consiglio Nazionale Ricerche. – 308 p.
Seventh Symposium on Mathematical Geodesy (4th Hotine Symposium); Proceedings; Assisi, Juni 8th to Juni 10th, 1978 (ENG) + (FRE)
AS 12.15.05/12.01.17
Key words: 4th Hotine symposium, mathematical geodesy, proceedings

Contents: Part 2

Livieratos, E. – On some structural characteristics of the mechanical and geometrical properties of the gravity field of the Earth in terms of the disturbing potential (p. 1–12)

Koch, K. R. – Parameter estimation in the Gauss-Helmert model (p. 13–23)

Baetsle, P. L. – Nouvelles formules pour les representation cylindriques conformes transverses de l'ellipsoide (Gauss-Krueger, U.T.M. etc.) (p. 25–34)

Bartelme, N.; Meissl, P. – Application of Toeplitz forms to the mathematical analysis of geodetic networks (p. 35–45)

Boucher, C. – The general theory of deformations and its applications in geodesy (p. 47–69)

Freeden, W.; Kersten, H. – Best approximate integration formulas for boundary surfaces of star-shaped bodies (p. 71–82)

Moritz, H.; Sanso, F. – A dialogue on collocation (p. 83–85)

Grossman, N. – Use of inertial navigation systems for determination of absolut height (p. 87–101)

Techn. Univ. Dresden IGD 84/I 303

BIBLIOGRAPHIA GEODAETICA

Authors

(The figures behind the authors indicate the number of the abstract; the Roman numerals indicate the group of the abstract)
Reprint only with permission of the editorship

No. 3/1984

061.3(100)+45"1978":528.23

. . .

VII Simposio sulla Geodesia Matematica; Assisi, 8—10 Giugno 1978
Firenze, 1981. — Commiss. Geod. Italiana, Gruppo Nazionale Geofis., Terra Solida del Consiglio Nazionale Ricerche. — 308 p.
Seventh Symposium on Mathematical Geodesy (4th Hotine Symposium); Proceedings; Assisi, Juni 8th to Juni 10th, 1978 (ENG)
AS 12.15.05/12.01.17
Key words: 4th Hotine symposium, mathematical geodesy, proceedings

Contents: Part 3

Hech, B. — Computation of second derivatives of the disturbing potential by surface layer methods (p. 103—118)

Sanso, F. — The minimum mean square estimation error principle in physical geodesy (stochastic and non-stochastic interpretation) (p. 119—137)

Moritz, H. — Statistical foundations of collocation (p. 139—194)

Suenkel, H. — Covariance approximations (p. 195—214)

Sanso, F.; Tscherning, C. C. — Notes on convergence in collocation theory (p. 215—246)

Witsch, K. J. — A uniqueness theorem for the geodetic boundary problem (p. 247—254)

Groten, E. — Geodesy-in-the large vs. differential geodesy (p. 255—262)

Krarup, T. — A convergence problem in collocation theory (p. 263—278)

Mather, R. S. — The Earth's gravity field and ocean dynamics (p. 279—308).

Techn. Univ. Dresden IGD 84/I 304

061.3(100)+45"1981":528.23

. . .

VIII Simposio sulla Geodesia Matematica; Como, 7—9 Settembre 1981
Firenze, 1983. — Comm. C.N.R. — Gruppo Nazionale Fisica Matem. et Geofis. Terra Solida. — Vol. I p. 1—415
Eighth Symposium on Mathematical Geodesy (5th Hotine Symposium); Proceedings; Como, September 7th to September 9th, 1981 (ENG/FRE)
AS 12.15.05/12.01.17
Key words: 5th Hotine symposium, mathematical geodesy, proceedings

Contents: Part 1

Dufour, H. M. — La correction de Rudski ameliorée — justifications — applications (p. 1—19)

Bode, A.; Grafarend, E. W. — The telluroid mapping based on a normal gravity potential including the centrifugal term (p. 21—56)

Chovitz, B. H. — The influence of Hotine's mathematical geodesy (p. 57—64)

Sanso, F.; Dermanis, A. — A geodynamic boundary value problem (p. 65—87)

Pellinen, L. P. — Effects of Earth ellipticity on solving geodetic boundary value problem (p. 89—103)

Freeden, W. — Interpolation and best approximation by harmonic spline functions — theoretical and computational aspects (p. 105—120)

Rummel, R. — Gravity parameter estimation from large and densely spaced homogeneous data sets (p. 121—132)

Techn. Univ. Dresden IGD 84/I 305

061.3(100)+45"1981":528.23

. . .
VIII Simposio sulla Geodesia Matematica; Como, 7—9 Settembre 1981
Firenze, 1983. — Comm. C.N.R. — Gruppo Nazionale Fisica Matem. et Geofis. Terra Solida. — Vol. I p. 1—415
Eighth Symposium on Mathematical Geodesy (5th Hotine Symposium); Proceedings; Como, September 7th to September 9th, 1981 (ENG/FRE)
AS 12.15.05/12.01.17
Key words: 5th Hotine symposium, mathematical geodesy, proceedings

Contents: Part 2

Caputo, M. — The estimate of the stress tensor on the Earth's surface (p. 133—147)

Petrovskaja, M. S. — Representation of the Earth's potential in form of convergent series (p. 149—156)

Krynski, J. — A method of global estimation of the gravity field from satellite-to-satellite data (p. 157—166)

Bosch, W. — Some remarks to the geopotential, its boundary values and the topography of the Earth (p. 167—178)

Fritsch, D.; Schaffrin, B. — The "choice of norm" problem for the free net adjustment with orientation parameters (p. 179—202)

Adam, J. — On the determination of similarity co-ordinate transformation parameters (p. 203—210)

Engl, H. W. — On least-squares collocation for solving linear integral equations of the first kind with noisy right-hand side (p. 211—233)

Techn. Univ. Dresden IGD 84/I 306

061.3(100)+45"1981":528.23

. . .
VIII Simposio sulla Geodesia Matematica; Como, 7—9 Settembre 1981
Firenze, 1983. — Comm. C.N.R. — Gruppo Nazionale Fisica Matem. et Geofis. Terra Solida. — Vol. I p. 1—415
Eighth Symposium on Mathematical Geodesy (5th Hotine Symposium); Proceedings; Como, September 7th to September 9th, 1981 (ENG/FRE)
AS 12.15.05/12.01.17
Key words: 5th Hotine symposium, mathematical geodesy, proceedings

Contents: Part 3

Kelm, R. — Modelling of local gravity field quantities for the determination of geometric parameters in local networks (p. 235—245)

Hofmann-Wellenhof, B. — Application of sparse matrix techniques to physical geodesy (p. 247—261)

Tscherning, C. C.; Poder, K. — Some geodetic applications of Clenshaw summation (p. 263—289)

Sjoeberg, L. — On the altimetry-gravimetry boundary problem (p. 291—306)

Eeg, J. — Continuous methods in least-squares theory (p. 307—321)

Nagy, D. — Direct gravity formulae for the geodetic reference systems 1980 (p. 323—334)

Sanso, F. — An note on density problems and the Runge-Krarup theorem (p. 335—361)

Fell, P.; Karaska, M. — Kernel function optimization for decreasing truncation error for the Stokes and Vening Meinesz integrals (p. 363—413)

Bertotti, B. — Statistics of narrow structures of the gravity field of the Earth. I. general theory (p. 415).

Techn. Univ. Dresden IGD 84/I 307

061.3(100)+45"1981":528.23

. . .

VIII Simposio sulla Geodesia Matematica; Como, 7—9 Settembre 1981
Firenze, 1983. — Comm. C.N.R. — Gruppo Nazionale Fisica Matem. et Geofis. Terra Solida. — Vol. II p. 417—640
Eighth Symposium on Mathematical Geodesy (5th Hotine Symposium); Proceedings; Como, September 7th to September 9th, 1981 (ENG)
AS 12.15.05/12.01.17
Key words: 5th Hotine symposium, mathematical geodesy, proceedings

Contents: Part 4

Bocchio, F. — On a gravity field with geodetic singularities (p. 417—428)

Holota, P. — The altimetry-gravimetry boundary value problem I: linearization, Friedrich's inequality (p. 429—448)

Svensson, S. L. — A new geodesy based upon inversion free Bjerhammar predictors (p. 449—468)

Dermanis, A.; Livieratos, E. — Dilatation, shear, rotation and energy analysis of map projections (p. 469—484)

Holota, P. — The altimetry-gravimetry boundary value problem II: weak solution, V-ellipticity (p. 485—500)

Techn. Univ. Dresden IGD 84/I 308

061.3(100)+45"1981":528.23

. . .

VIII Simposio sulla Geodesia Matematica; Como, 7—9 Settembre 1981
Firenze, 1983. — Comm. C.N.R. — Gruppo Nazionale Fisica Matem. et Geofis. Terra Solida. — Vol. II p. 417—640
Eighth Symposium on Mathematical Geodesy (5th Hotine Symposium); Proceedings; Como, September 7th to September 9th, 1981 (ENG)
AS 12.15.05/12.01.17
Key words: 5th Hotine symposium, mathematical geodesy, proceedings

Contents: Part 5

Lachapelle, G.; Mainville, A.; Schwarz, K.-P. — Empirical investigations on the anisotropy of the free air gravity anomaly covariance function (p. 501—515)

Bulirsch, R.; Gerstl, M. — Numerical evaluation of elliptic integrals for geodetic applications (p. 517—528)

Georgiadou, Y.; Grafarend, E. W. — The small scale structure of geometry and gravity space I (p. 529—555)

Remmer, O. — Two letters on physical geodesy (p. 557—588)

Colombo, O. L. — The convergence of the external spherical harmonic expansions of the gravitational potential (p. 589—607)

Weightman, J. A. — Singular cases of the geometric significance of combining geodetic net adjustments (p. 609—623)

Brunner, F. K. — Appraisal of the geometrical-optics approximation for geodetic refraction calculations (p. 625).

Special Study Group 4.56 — "Differential geometry of the gravity field", by Grafarend, E. W.

Special Study Group 4.57 — Report of the activities, by Sanso, F.

Study Group 4.58 — "Representation of the gravity field", by Dufour, H. M.

Techn. Univ. Dresden IGD 84/I 309

AIDOS Systematics **12.15.09** **Group I Geodesy**

Physical geodesy

Theory and methods for exploring the figure and the gravity field of the Earth
Constants of the Earth, physics of the Earth
Movements of the Earth's crust, geotectonics
Geodynamic investigations, gravity measurements
Gravimetry for potential-theoretical foundations

Gruppe I Höhere Geodäsie

Physikalische Geodäsie

Theorie und Methoden zur Erforschung der Figur und des Schwerefeldes der Erde
Konstanten der Erde, Physik der Erde
Erdkrustenbewegungen, Geotektonik
Geodynamische Untersuchungen, Schweremessungen
Gravimetrie für potentialtheoretische Grundlagen

528.241(495)

Arabelos, D. / Tziavos, I. N.
Determination of deflections of the vertical using a combination of spherical harmonics and gravimetric data for the area of Greece (ENG)
Bull. geod. — Paris 57 (1983) 3. — p. 240—256 : 6 fig. 4 tab. 27 lit.
AS 12.15.09
Key words: Deflection of the vertical, Greece

The determination of gravimetric deflections of the vertical for the area of Greece is attempted by combining a spherical harmonics model and gravity anomalies using the method of least squares collocation. The components of deflections of the vertical are estimated on a grid with spacing 15′ in latitude and 20′ in longitude covering only the continental area of Greece, where a sufficient number of point gravity anomalies is available. In order to test the accuracy of the determination, gravimetric deflections of the vertical are computed at stations where astrogeodetic data are available. The results show that in a large region of rugged topography and irregular potential field, the prediction is possible with a standard deviation of 18 % ... 28 % of the root mean square variation of the observations, without taking into account the topography. Furthermore, the estimation of some systematic differences between observed and computed deflections of the vertical is attempted.
(Author's summ.)

Techn. Univ. Dresden IGD 84/I 310

061.24(100):528.27

Balmino, G.
A few remarks on the Earth's atmosphere effects on the geoid, the free-air gravity anomalies, the altimetric geoid, and combination procedures in global geopotential model computation (ENG)
Bull. Inform. — Toulouse (1983) 53. — p. 219—232 : 2 fig. 9 lit.
AS 12.15.09
Key words: Earth's atmosphere, geoid, geopotential, gravity anomaly

This is a synthetic note on the various effects of the Earth's atmosphere on the definition (and computation) of some geodetic quantities. A particular emphasis is put on the continuity aspect of formulas describing the Earth's gravity field at satellite altitudes and at the Earth's surface. Spherical approximations are used throughout.
(Author's summ.)

Techn. Univ. Dresden IGD 84/I 311

528.223

Biro, Peter
Physikalische Interpretation der Messergebnisse von wiederholten geodaetischen Beobachtungen
Vermessungstechnik. — Berlin 32 (1984) 2. — p. 48—49 : 8 lit.
Physical interpretation of measuring results of repeated geodetic observations (GER)
AS 12.15.09/12.15.33
Key words: Earth's gravity field, physical geodesy

Geodetic observations are functions of position and of the Earth's gravity field. Previously, the Earth body was assumed to be rigid and the gravity field to be constant (statistical conception). Later on, geodesy took cognizance of deformations of the Earth body and developed a semi-dynamic mode of consideration. A consistent, fully dynamic model for the interpretation of measuring results of repeated observations may be created with "integrated" geodesy.

(Author's summ.)

Techn. Univ. Dresden IGD 84/I 312

528.27

Brovar, V. V.
Vlijanie peremennojj atmosfery na gravitacionnoe pole Zemli
Izuch. Zemli kak planety metodami astronomii, geod. i geofiz. Tr. 1 Orlov. konf., Kiev 29 sent. — 3 okt. 1980. — Kiev (1982). — p. 92—93
The effect of the variable atmosphere on the gravity field of the Earth (RUS)
AS 12.15.09/49.21
Key words: Atmosphere, gravity field

Atmospheric density variations cause variable loads on the surface of land, sea and oceanic bottom. It is thus necessary to take into account both the attraction of anomalous atmospheric masses and the deformations due to them. The measured gravity value should be improved by applying the corrections as follow. The first correction is due to the direct effect and may amount some tens of mkGal. The second one accounts for attraction of those anomalous masses over the sea surface the pressure of which is not yet compensated by water redistribution. The redistribution depends on roughness of the bottom, shore line shape and the preceding state of the atmosphere. The third correction accounts for elastic deformations of the land and bottom surfaces.

Dobrokhotov, National Correspondent, USSR IGD 84/I 313

528.27:629.783

Chapanov, Javor / Georgiev, Nicola
Opredelenie gravitaciannykh vozmushhenijj v promezhutochnam dvizheni ISZ
Vissha Geod. — Sofija (1983) 9. — p. 3—7
Determination of the gravity perturbations in intermediate motion of artificial Earth satellites (BUL)
AS 12.15.09/12.15.17
Key words: Gravity measurement, satellite geodesy

Power series by regularized time are used to determine, on the basis of the generalized problem of two fixed centres, the perturbations in the intermediate motion of artificial Earth satellites due to zonal harmonics of the Earth gravity potential. The formulae deduced for the coefficients of the power series development of perturbations provide an accuracy of 10^{-9} of the coordinates of artificial Earth satellite which corresponds to the use of orbital methods in geodesy.

(Author's summ.)

Georgiev, National Correspondent of Bulgaria IGD 84/I 314

528.375:528.27

Embacher, W.
Ueber den Einfluß der Schwere auf die Hoehe
Vermess. Photogramm. Kulturtechnik. – Zuerich 81 (1983) 10. – p. 354–359 : 3 fig. 5 lit.
The effect of gravity on height (GER)
AS 12.15.09/12.15.33
Key words: Gravity measurement, height determination, levelling

Trigonometrical as well as geometrical levellings provide differences in height between a mathematically defined reference surface through the initial point and the corresponding parallel surface through the target point or the terminal. Dynamic levelling, i.e. the determination of differences in height from gravity differences, also refers to a mathematically and physically defined surface on which the theoretical gravity values γ_0 are known. This normal gravity γ_0 may be determined by tables with the latitude φ of the measuring point and provides the difference $(g - \gamma_0)$. By measurement of the second gravity value with a difference in height Δh to the measuring point we obtain the differential quotient $\frac{\Delta g}{\Delta h}$. The determination of the inclination of terrain δ in the line of greatest slope and the surface density σ is also possible.

(Author's summ.)

Techn. Univ. Dresden IGD 84/I 315

528.38.024.1:551.241(084.3)

Foeldvary-Varga, M.
Investigation by levelling errors of closure of reliability of velocity values of the recent vertical crustal movements of Hungary (ENG)
Period. Polytechn. : Ser. Civil Engin. – Budapest 27 (1983) 3–4. – p. 189–196 : 3 fig. 2 tab. 10 lit.
AS 12.15.09/12.15.37/12.15.33
Key words: Error of closure, levelling net, vertical crustal movement

A geodetic method is applied to investigate the reliability of the Hungarian part of the international vertical crustal movements maps prepared until now. The accuracy is investigated by reducing the measurements of levelling lines measured at different times to a chosen epoch by velocity values read off the movements maps, then polygons are put together from reduced values and their errors of closure computed. Reliability of velocity maps is judged from the changes of original errors of closure.

Veroe-Hetenyi, National Correspondent, Hungary IGD 84/I 316

528.223:528.1

Freeden, Willi
Least square approximation by linear combinations of (multi)poles (ENG)
Columbus: Ohio State Univ., Rep. Dep. Geod. Sci. & Surv. (1983) 344. – 86 p. : div. fig. div. lit.
AS 12.15.09/12.15.37
Key words: Functional analysis, least square approximation, potential theory

The purpose of the paper is to transcribe the standard procedure of expanding an external gravitational potential by trial functions (e.g., mass points, multipoles, spherical harmonics) from the classical spherical case to nonspherical models.

The external gravitational potential of the Earth is shown to be expandable into a series convergent outside and on any closed surface completely surrounding the Earth's surface in the outer space. In addition, for each compact subset of the outer space with positive distance to the Earth's surface, the convergence is uniform.

(Author's summ.)

Techn. Univ. Dresden IGD 84/I 317

528.02:551.241

Fujii, Yoichiro / Nakane, Katsumi

Horizontal crustal movements in the Kanto-Tokai district, Japan, as deduced from geodetic data (ENG)

Tectonophysics. — Amsterdam (1983) 97. — p. 115—140 : 12 fig. 1 tab. 43 lit.

AS 12.15.09/12.15.33/49.31

Key words: Earthquake, Earth's crust movement

The Kanto and Tokai district, Japan, is located very close to the northern convergent boundary of the Philippine Sea plate and has been attacked by great earthquakes several times during the historic past. This district has been resurveyed several times since the first observation. The authors compiled all the geodetic data and deduced the horizontal crustal movements. In the new net-adjustment for geodetic network, the six stations, Tsukubasan, Teruishizan, Dodairayama, Kokushidake, Kaikomagadake, and Enasan are assumed not to have moved during the observational period. The horizontal displacement vectors for four periods are deduced. Also error ellipses of displacement vectors are estimated. The Earth's crustal strains are calculated from the obtained vectors.

Nagasawa, National Correspondent, Japan IGD 84/I 318

528.21:629.783

Gajazov, I. S.

Ispol'zovanie modeli geopotenciala v orbital'nom metode opredelenija koordinat stancijj

Bjull. In-ta teor. astron. AN SSSR. — Moskva (1982) 4. — p. 200—204 : 11 lit.

The use of geopotential model in the orbital method of station positioning (RUS)

AS 12.15.09/12.15.17

Key words: Geopotential, positioning, satellite

From an analysis of spherical co-ordinate errors of stations on the Earth's surface by the orbital method it is established that the conditions for determination of the radius-vectors of stations are seldom optimal and cause thus considerable errors of this co-ordinate in the adjustment. Therefore, it is suggested to use as measurements the values of radii-vector computed for the so-called "planetary geoid" assuming an apriori gravitational potential on its surface with accounting for the height of the station on the physical surface of the Earth above the mean sea level as determined from levelling. The error of the radius-vector computed in such a way will be 4 to 5 m when taken all over the globe.

Dobrokhotov, National Correspondent, USSR IGD 84/I 319

528.247:550.34

Hagiwara, Yukio

Seismic intensity of the 1982 Urakawa-Oki earthquake as related to the Bouguer gravity anomaly (JAP)

J. Geod. Soc. Japan. — Tokyo 29 (1983) 3. — p. 164—165 : 3 fig. 4 lit.

AS 12.15.09

Key words: Earthquake, gravity anomaly

Each time a large earthquake occurs in Japan, a detailed survey of seismic intensity is made by a questionnaire on inhabitants in the seismic area. In a rough estimate, seismic intensity increases with magnitude and decreases with hypocentric distance. The standard seismic intensity (JMA scale) is empirically determined by Kawasumi's formula as a function of magnitude and hypocentric distance.

The seismic intensity anomaly is defined by the deviation of an actually observed intensity from the standard one. On the basis of the results of the survey made immediately after the 1982 Urakawa-Oki earthquake of M7.1, it becomes clear that the distribution of seismic intensity anomaly agrees well with that of long-wavelength Bouguer anomaly.

Nagasawa, National Correspondent, Japan IGD 84/I 320

528.27:550.38

Hashiguchi, Masamichi et al.
Observation of precise geomagnetic anomalies in total force at Sagami Bay (JAP)
J. Geod. Soc. Japan. – Tokyo 29 (1983) 3. – p. 157–163 : 8 fig. 3 lit.
AS 12.15.09/49.31
Key words: Geomagnetism, physical geodesy

In order to know precise geomagnetic anomalies in total force at Sagami Bay where the large Kanto earthquake occurred in 1923, magnetic survey was carried out by use of towing proton-precession magnetometers.

The effect of magnetization of the ship was removed by use of a pair of magnetometers towed 50 m and 200 m apart from stern of the ship. Ship's position was determined by use of Loran C within an accuracy of 50 m. Magnetic daily variation was removed by use of data given from the ocean bottom magnetometer deployed at the same region.

By repeating the observation, it would be possible to detect variation of local geomagnetic anomalies due to earthquakes or volcanic activity.

Nagasawa, National Correspondent, Japan IGD 84/I 321

(528.28:629.783):621.396.962
521.3/.4:629.783

Hauck, Hermann
The program ORBDOP (Extended GEODOP) (ENG)
Nachr. Karten- u. Vermess.-Wesen, R. II. – Frankfurt/M. (1983) 40. – 8 fig. 5 lit.
AS 12.15.09/12.15.17/12.15.33
Key words: Doppler measurement, Doppler program GEODOP, satellite, short arc program ORBDOP

The Canadian semi short arc Doppler program GEODOP has been extended to the short arc program ORBDOP.

In ORBDOP no Precise Ephemerides are needed. The orbit is computed by the integration of the disturbing equation system using an approximative starting state vector at a defined epoch. The integration is based on Hill-variables.

It is shown how ORBDOP is constructed and how to handle the program.

(Author's summ.)

Techn. Univ. Dresden IGD 84/I 322

528.2

Izotov, A. A.
Some problems of geodynamics and methods for their solution (ENG)
Acta Geodaet., Geophys. et Montan. – Budapest 18 (1983) 1–2. – p. 9–14
AS 12.15.09/12.15.17/12.15.13
Key words: Crustal movements, geodynamics, geokinetics, physical geodesy, plate tectonics, polar motion, satellite geodesy, spherical astronomic

Inner and outer forces influencing the present evolution of the Earth can be detected and interpreted by the effects caused in the Earth's crust and surface (change of the rotation velocity, polar motion, Earth tides, vertical and horizontal crustal movements, movements of lithospheric plates). The most urgent problems in this field are the connection of physico-chemical processes in the Earth's interior with tectonic phenomena, the interaction between continental and oceanic parts of the crust, and the development of oceanic trenches and their effect on the crustal evolution, present and past continental movements.

In the framework of the international geodynamics project the problems are approached from a geological-geophysical point of view. These methods, however, cannot yield accurate numerical data on the movements. Global geokinetic processes could be studied most easily by satellite geodetic and astronomic methods. The author discusses such methods and proposes to establish a mixed commission of astronomers-geodesists and geologists-geophysicists which should outline the problems to be solved by common efforts.

Veroe-Hetenyi, National Correspondent, Hungary IGD 84/I 323

550.312:551.241

Kurbanov, M. K. / Kirsta, O. B. / Gaipov, B. N.
Neprilivnye variacii sily tjazhesti kak rezul'tat izbytochnykh naprjazhenijj v zemnojj kore
Izvest. AN TSSR. Ser. fiz.-tekhn. khim. i geol. n. — Dushanbe/SU (1981) 4. — p. 125—127 : 5 lit.
Non-tidal gravity variations as the result of surplus tensions in the Earth's crust (RUS)
AS 12.15.09
Key words: Earth's crust, gravity observation

Dobrokhotov, National Correspondent, USSR IGD 84/I 324

525.1:528.21

Marchenko, A. N.
O nekotorykh fundamental'nykh postojannykh, opredeljajushhikh dinamicheskuju i geometricheskuju figury Zemli i planet
Izuch. Zemli kak planety metodami astronom. geod. i geof. Tr. I Orlov. konf., Kiev 29 sept. — 3 okt., 1980. — Kiev (1982). — p. 11—16 : 12 lit.
On some fundamental constants determining dynamic and geometric figures of the Earth and the planets (RUS)
AS 12.15.09/12.15.13
Key words: Earth's figure, fundamental constants, planets

The formulas that allow determination of geometrical parameters of an equipotential surface corresponding to the spherical function of the 2nd order of the gravitational potential of a planet from its Maxwell parameters are given. The Maxwell parameters for potential models of Earth, Moon, Venus, Mars, and Jupiter are computed and discussed.

Dobrokhotov, National Correspondent, USSR IGD 84/I 325

529.932(084.3-16)(480):528.247

Mikkola, Leena
Mean height map of Finland (ENG)
Suomen geod. Laitoksen Julkaisuja. — Helsinki (1983) 98. — 5 p. : 2 fig. 1 cart app.
AS 12.15.09
Key words: Gravity anomaly, mean height map

The main purpose of the mean height map of Finland was to calculate the terrain-corrected Bouguer anomalies. The whole country was divided into squares measuring 10 · 10 qkm in the Gauss-Krueger plane co-ordinate system for zone 27°. For each mean height 16—121 spot heights were interpolated on large-scale maps.

Each mean height refers only to that part of the square which is within the borders of Finland.

(Author's summ.)

Techn. Univ. Dresden IGD 84/I 326

528.27:528.563.089.6

Nakagawa, Ichiro et al.
International gravimetric connections along the Circum-Pacific zone (IV) — Drift characteristics of LaCoste and Romberg gravimeters (model G) — (JAP)
J. Geod. Soc. Japan. — Tokyo 29 (1983) 3. — p. 141—149 : 5 fig. 4 lit.
AS 12.15.09/73.19
Key words: Drift of gravimeter, gravimeter, physical geodesy

A method how to accurately determine the drift of gravimeter is one of important but difficult problems in the analysis of gravimetric data, especially for data obtained by international gravimetric connections which were carried out taking a long duration of time. The drift of LaCoste and Romberg gravimeters (model G) was thoroughly examined, using the data obtained by the international gravimetric connections which were carried out at selected cities mainly along the Circum-Pacific zone from 1979 to 1982.

It was clearly recognized for some gravimeters that positive and negative drifts of the gravimeter occurred when the gravimeter was brought from cities with large and small gravity values, respectively. Generally speaking, however, cities with large gravity values are cooler, while those with small values are warmer. Such a pronounced tendency on the drift of the gravimeter may be due to an elastic property of the springs mounted inside of the gravimeter.

Nagasawa, National Correspondent, Japan IGD 84/I 327

528.27:528.563.089.6

Nakagawa, Ichiro et al.
International gravimetric connections along the Circum-Pacific zone (V) — Effects of the rapid change of atmospheric pressure upon the reading values of LaCoste and Romberg gravimeters (model G) — (JAP)
J. Geod. Soc. Japan. — Tokyo 29 (1983) 3. — p. 150—156 : 5 fig. 3 lit.
AS 12.15.09/73.19
Key words: Atmospheric pressure, gravimeter, gravity measurement, physical geodes

In precise gravity surveys of mountainous areas, a sudden and large change in reading values of LaCoste and Romberg gravimeters has frequently been experienced immediately after climbing down from a high altitude.

In the course of international gravimetric connections, gravity measurements in mountainous areas were performed by means of several LaCoste and Romberg gravimeters (model G) along a route from Valparaiso on the Pacific coast to Caracoles on the border of Chile-Argentina in Chile. This route was satisfactory for exmining the effects of the rapid change of atmospheric pressure on the gravimeter readings. Because the measuring stations are located at a height difference amounting to more than 3,000 m, and it was of great advantage to the examination that it took only a single day by car to travel to and from both the measuring stations.

It was ascertained that relative deviations in reading values for three gravimeters employed were well correlated with the difference of station altitudes.

Nagasawa, National Correspondent, Japan IGD 84/I 328

528.27:525.6

Ogier, M. / Sakuma, A.
Variations de la pesanteur non liees a la maree luni-solaire
Bull. Inform. — Toulouse (1983) 53. — p. 129—137 : 7 fig. 3 tab.
Gravity variations not dependent on luni-solar tides (FRE)
AS 12.15.09/12.15.21/73.19
Key words: Gravimeter, gravity variation

The measurement in France from 1983 January to July of 8 absolute gravity values at 6 stations allow us to make some investigations for non-tidal gravity variations.

The instrumental accuracy of the absolute gravimeter GA 60 from Jaeger being about 4 μGal (r.m.s.), we can find and determine with accuracy for two stations a correlation between gravity changes and barometric pressure variations. The computed coefficients were:

— 4.28 et 4.78 nm/s/s per mBar for Orleans

— 4.4 nm/s/s per mBar for Toulouse.

In addition, two measurements performed in Orleans at 6 months interval (1983 February—July) don't show any gravity change despite a 0.3 m change of the water table.

(Author's summ.)

Techn. Univ. Dresden IGD 84/I 329

525.6:531.781.3

Ozawa, Izuo
Observations of the vertical strains of the Earth tide in the tunnel (JAP)
Disaster Prevention Res. Inst. Ann. — 26 (1983) B-1. — p. 77—86 : 5 fig. 1 tab. 6 lit.
AS 12.15.09/73.19
Key words: Earth tide, extensometer observation, physical geodesy, vertical strain measurement

Observations of the vertical strains of the Earth tide have been made on three kinds of vertical extensions in the former Osakayama railway tunnel since 1952.

The extensometers have been set in the well, on the vertical radius and on the side wall in the tunnel respectively.

The tidal analyses have been made for the data of the three kinds of extensometers.

The relations between the tidal components and the numbers of the days of the analysed periods have been analysed.

The effects of the tunnel on the tidal extensions on the vertical chords and the depth of the well in the tunnel have also been studied.

Nagasawa, National Correspondent, Japan IGD 84/I 330

525.6:528.541.4

Ozawa, Izuo / Fujii, Kazunari
Observations of the Earth tide by means of strain type tiltmeters (rotationmeters) (JAP)
J. Seism. Soc. Japan, II. — 36 (1983) 12. — p. 195—202 : 3 fig. 1 tab. div. lit.
AS 12.15.09/73.19
Key words: Earth tide, physical geodesy, tiltmeter

A strain type tiltmeter (rotationmeter) which is able to observe the rotational strain with respect to the horizontal axis has been devised. These instruments have been set in the directions of N38°E and N52°W in the Osakayama tunnel in 1978. They have stable observational recordings at a sensitivity of around 0.″0027/mm. And their observations have been performed for long periods very well.

Their Earth tidal components have been analyzed on the observed records for the periods of three months. These results show that the amplitude ratios between the components tides have good fits to those of the direct terms of the tide generating potential at the observation places in the direction of N52°W (rotational strain with respect to N38°E axis) but the S_2-component is predominant in the direction of N38°E (rotational strain with respect to N52°W axis). Then, the vertical deflection of the crust is calculated by means of combinations with the results of the tidal analyses of the horizontal pendulum type tiltmeters and water-tube tiltmeter.

Nagasawa, National Correspondent, Japan IGD 84/I 331

528.247:519.218.82

Paul, M. K.
Orientation-independent filtering in gravity analysis (ENG)
Acta geod. geophys. montan. Acad. Sci. Hung. – Budapest 16 (1981) 1. – p. 63–72 : 2 fig. 4 tab. 15 lit.
AS 12.15.09/12.15.33/12.15.37
Key words: Filtering, gravity anomaly

Some formulae for the coefficients for band-pass filtering of Bouguer gravity anomaly have been derived so that when they are used in the weighted summations of gridded gravity values, the results are independent of the orientation of the axes of rectangular grids. Such a derivation has been achieved by starting from a transfer function of circular symmetry that fulfils the requirements of a band-pass filter and then integrating its contribution over a rectangle with a grid point at its center. Computational checks applied to different sets of coefficients as evaluated by the proposed method confirm their accuracy to at least five significant digits.

(Author's summ.)

Techn. Univ. Dresden IGD 84/I 332

528.2(47+57):001.8

Pellinen, L. P.
Sovremennye raboty CNIIGAiK v oblasti fizicheskojj geodezii
Geod. i Kartogr. – Moskva (1983) 1. – p. 30–36 : 8 lit.
Recent research of the Central Research Institute of Geodesy, Air Survey and Cartography (ZNIIGAiK) in the field of physical geodesy (RUS)
AS 12.15.09
Key words: Physical geodesy, research, USSR

The main trends of ZNIIGAiK research in the field of physical geodesy were formed in 1930–1940. In recent years regional studies of the figure of the Earth are carried out for determination of quasigeoidal heights and deflections of the vertical to use them in the forthcoming adjustment of the national net of the USSR. The improvements of the astrogravimetric levelling were able to reduce considerably the effect of random errors. A dense net of levelling loops using astronomical stations in the first- and second-order geodetic nets has been established. The effect of ellipticity of the Earth in solution of geodetic problems has been investigated. A new method of approximation of the external gravity field of the Earth by point masses using the gravity anomalies having the most clear relation with point mass parameters was proposed.

Dobrokhotov, National Correspondent, USSR IGD 84/I 333

528.223:523.3/.4

Sagitov, M. U.
Gravitacionnye polja Zemli, Luny i planet i ikh sravnitel'nyjj analiz
Izuch. Zamli kak planety metodami astronomii, geod. i geofiz. Tr. 1 Orlov. konf., Kiev 29 sent. – 3 okt. 1980. – Kiev (1982). – p. 63–78 : 32 lit.
Gravitational fields of Earth, Moon and the planets and their comparative analysis (RUS)
AS 12.15.09/12.15.13
Key words: Earth, gravity field, Moon, planets

The accuracy of determination of harmonics of the geopotential from combination of terrestrial and satellite data is estimated. The gravimetric coverage of Earth, Moon, Mars, Phobos, Venus and Jupiter as well as the global and local features of the fields are discussed. A pair similarity of the general variation of degree variances for Earth and Venus resp. Moon and Mars that seems to be connected with cosmogonic history has been revealed. The mean gravity on Phobos is 0.5 Gal while the anomalies vary within 70 mGal. The gravitational field of small bodies should be studied by means of a gradient-meter. The dificulties of computation of the second vertical gravity gradient are stressed.

Dobrokhotov, National Correspondent, USSR IGD 84/I 334

528.21(624):621.396.962

Salih, A. B.
The shape of the geoid in the Sudan (ENG)
Austral. J. Geod., Photogramm. & Surv. – Sydney (1983) 39. – p. 27–38
AS 12.15.09
Key words: Doppler, geoid

The geoid shape in the Sudan is determined by two methods: a) the astrogeodetic, and b) the dynamic methods. The data for a) has accumulated since 1937. The data for b) is from observations made in the years 1975, 1978 and 1981 using a Doppler receiver (JMR). A comparison of the two geoids shows good agreement with each other but they clearly show a misfit from east to west with the local geodetic datum (Adindan), based on the Clarke 1880 Spheroid. A better fit is shown in the north-south direction.

Gilliland, National Correspondent of Australia IGD 84/I 335

528.28:629.783
621.396.962:528.02

Schlueter, Wolfgang
Satellite Doppler point positioning using the Navy Navigation Satellite System (ENG)
Manuscript prepared for a series of lectures held at the Wuhan College of Geodesy, Cartography and Photogrammetry, Nov.–Dec. 1981
Nachr. Karten- u. Vermess.-Wesen, R. II. – Frankfurt/M. (1983) 40. – p. 23–97 : 24 fig. 13 tab. 13 lit.
AS 12.15.09/12.15.17
Key words: Doppler measurement, positioning, satellite system

The technique of Doppler point positioning using the U.S. Navy Navigation Satellite System is presented. A description of the Navy Navigation Satellite System (NNSS) itself and of some geodetic receiver including the data output is given. The formulae for single point positioning and for (semi) short arc solutions are derived and refraction models are discussed. Finally the data reduction concept using the program GEODOP is outlined.

(Author's summ.)

Techn. Univ. Dresden IGD 84/I 336

528.28:629.783
621.396.962:528.02

Schlueter, Wolfgang / Amberg, Ludwig
Final results of Doppler point positioning at several timekeeping laboratories in Europe (ENG)
Nachr. Karten- u. Vermess.-Wesen, R. II. – Frankfurt/M. (1983) 40. – p. 99–107 : 2 tab. 2 lit.
AS 12.15.09/12.15.17
Key words: Doppler measurement, positioning, satellite system

At 21 stations in Europe the point positioning was carried out by using the Doppler technique. The observations of the satellites of the U.S. Navy Navigation Satellite System (NNSS) have been performed in the period from 6th to 16th November 1979. The co-ordinates computed with the precise ephemeris are given in this paper.
(Author's summ.)
Techn. Univ. Dresden IGD 84/I 337

528.021.6-187.4

Shanurov, G. A.
Voprosy obrabotki i ocenki tochnosti rezul'tatov radiointerferometrii so sverkhdlinnojj bazojj v edinojj sisteme koordinat
Izvest. vyssh. uchebn. Zaved., Geod. i Aehrofoto"jomka. – Moskva (1982) 4. – p. 45–51 : 12 lit.
The problems of treatment and accuracy estimation of VLBI-data in a homogeneous co-ordinate system (RUS)
AS 12.15.09/12.15.17/73.19
Key words: Co-ordinate system, geodetic astronomy, VLBI-technique

The equations relating the values of time delay and interference frequency to be measured with the values to be determined are derived. Among the latter there are the co-ordinates of radio sources, polar co-ordinates, parameters of precession and nutation as well as the co-ordinates of the vector of the interferometric base in the co-ordinate system fixed for the initial epoch. A method of polar position determination from co-ordinate variations of base vectors in a momentary co-ordinate system is suggested. It allows utilization of only one base oriented in a definite way, and in the case of several bases available it reduces the effect of station deformations upon the accuracy of polar position determination. It is concluded that polar co-ordinates and base vectors may be determined in a homogeneous co-ordinate system with the R.M.S. error within 1.5 R.M.S. measuring error.
Dobrokhotov, National Correspondent, USSR IGD 84/I 338

528.21.088(437)

Simek, Jaroslav
Metody i tochnost' opredelenija detal'nojj figury kvazigeoida na territorii ChSSR
Sbornik vyzkumn. praci VUGTK, R. 3. – Praha (1982) 14. – p. 39–61 : 5 tab. 16 lit.
Methods of determination of the detailed quasigeoid on the territory of the CSSR and its accuracy (RUS)
AS 12.15.09
Key words: Geodetic astronomy, geoid, gravimetry, levelling

The detailed quasigeoid was investigated by means of astrogravimetric and astronomic levelling. Quasigeoidal height differences were determined along the sides of two nets. The first triangular net consists of 53 selected astrogeodetic points the accuracy of which was independently checked with the help of gravity data

and 127 sides of average length of 57 km, the second quadrilateral is created by a system of meridians and parallels with intervals of 10′ and 15′ and consists of about 500 points and over 800 sides. Gravimetric deflections and quasigeoidal heights with the astrogravimetric levelling were determined with the help of removing and additional restoring of topographical masses according to Pellinen. For astronomical levelling the vertical deflections were densified by gravimetric interpolation along meridional and parallel profiles up to the intervals 6 km and 4.5 km.
Parametric adjustment of both nets was performed in several variants.

(Author's summ., abbr.)

Techn. Univ. Dresden IGD 84/I 339

528.241:528.23

Stojjnov, Vladimir
Edin nachin za opredeljane na komponentite na otklonenieto na otvesa pri reshavane zadachata na Molodenski
Vissha Geod. — Sofija (1983) 9. — p. 73–78
A method to determine deflections of the vertical in the solution of the Molodensky problem (BUL)
AS 12.15.09/12.15.05
Key words: Deflections of the vertical, mathematical geodesy, physical geodesy

The gravimetric components of the deflections of the vertical at a point of the physical Earth surface are determined by means of integrals in the expressions under the signs of which participate the null-order approximations of the two horizontal components, the gravity anomaly and the relief elements. The formulae for the components are obtained from the Green's formula of the perturbing potential, by substituting the relief elements and the components of the deflections of the vertical for the derivatives of this potential in the horizontal direction. The elements of the curvature of the normal sections do not participate in the formulae obtained which is their main particularity, as compared to the known formulae deduced on the basis of the integral equation of perturbing potential.

(Author's summ.)

Georgiev, National Correspondent of Bulgaria IGD 84/I 340

528.21(4):528.24

Torge, Wolfgang / Weber, Georg / Wenzel, Hans-Georg
Ein hochaufloesendes gravimetrisches Geoid fuer Europa und angrenzende Meeresbereiche
Z. Vermess.-Wesen. — Stuttgart 108 (1983) 8. — p. 321–331 : 4 fig. 4 tab. 27 lit.
A high resolution gravimetric geoid covering the adjacent seas (GER)
AS 12.15.09
Key words: Comparing investigation, determination of the Earth's figure, error analysis, Europe, geodetic computation, geodetic reference surface, geoid, gravimetric geodesy, gravity anomaly, physical techniques, variance

A high resolution gravimetric geoid has been determined in the area $30° < \varphi < 73°$ and $-30° < \lambda < 46°$ in a 12′ by 20′ grid covering Europe, the North Sea, Baltic Sea, Mediterranean Sea and parts of the North Atlantic Ocean. The computation has been carried out by least squares spectral combination using closed integral formulas, combining 104,000 mean 6′ by 10′ free air gravity anomalies, 12,000 mean 1° by 1° free air gravity anomalies and the spherical harmonic model GEM 9 complete to degree and order 20. The precision of the computed geoid heights has been estimated to ± 1 m. The computed gravimetric geoid heights have been compared with geoid heights derived from Seasat-1 altimetry in the North Sea and Mediterranean Sea and with geoid heights computed from the spherical harmonic model RAPP81 complete to degree and order 180. After bias reduction, rms discrepancies between 0.1 m and 0.7 m for the comparison with Seasat-1 altimetry and between 0.6 m and 1.3 m for the comparison with the spherical harmonic model RAPP81 confirm the estimated precision of ± 1 m for the gravimetric geoid EGG1.

(Author's summ.)

Techn. Univ. Dresden IGD 84/I 341

528.247:550.312

Ushakov, S. A. / Lazarev, G. E. / Rukin, M. D.
Gravitacionnoe pole i neotektonika litosfery Antarktiki
Probl. neotek. i sovrem. dinam. litosfery. Tez. dokl. Vses. soveshh., Tallin, apr., 1982. — Tallin (1982) 64
The gravitational field and the neotectonics of lithosphere of the Antarctic (RUS)
AS 12.15.09
Key words: Gravity anomaly, gravity field, tectonics

The models of finite kinematics of lithospheric plates of the southern hemisphere from the beginning of Gondvana destruction up to now are discussed. The schemes of bottom relief evolution of South Atlantic and Indian oceans are proposed. Theoretical models for evolution of the relief and the anomalous gravity field in the transition zone from the Eastern Antarctic to the Atlantic and Indian oceans are constructed.

Dobrokhotov, National Correspondent, USSR IGD 84/I 342

528.024.1:551.51

Volkov, V. I.
O vlijanii anomal'nykh atmosfernykh mass i poverkhnostnogo temperaturnogo ehffekta na rezul'taty povtornogo nivelirovanija
Geod. kontrol' v str-ve. — Kujjbyshev (1982). — p. 21—37 : 16 lit.
On the effect of anomalous atmospheric masses and ground temperature upon repeated levelling (RUS)
AS 12.15.09/12.15.33
Key words: Atmosphere, levelling, temperature

The studies were carried out for 4 specially monumented groups of bench marks in 1980 to 1981. Every group contained a deep bench mark monumented at a depth of 17.2 to 18.7 m as well as a rock or ground checking bench mark monumented at a depth of 0.8—2.0 m. It is established that anomalous atmospheric loads cause elastic deformations of surface crustal layers and for pressure variations of up to 28 mm of Hg the results of repeated high precision levelling may be distorted up to 2—5 mm. Diurnal and seasonal temperature variations of the ground cause different vertical displacements of bench marks depending on physical mechanical structure of the rocks, vegetation, ground water level and other parameters.

Dobrokhotov, National Correspondent, USSR IGD 84/I 343

525.731(26)

Vshivkov, V. F. / Shilkin, P. A.
Opredelenie atmosfernojj refrakcii nad morskojj poverkhnost'ju
Geod. i Kartogr. — Moskva (1982) 11. — p. 50—51 : 2 lit.
Determination of atmospheric refraction over the sea surface (RUS)
AS 12.15.09/12.15.21
Key words: Marine geodesy, refraction

Dobrokhotov, National Correspondent, USSR IGD 84/I 344

528.38(437):551.513

Vyskocil, Pavel
Refraction in levelling (ENG)
Sbornik vyzkumn. praci VUGTK, R. 3. – Praha (1982) 14. – p. 63–87 : 9 tab. 13 lit.
AS 12.15.09/12.15.33/12.15.25
Key words: Levelling, refraction

The results of investigation concerning the influence of refraction on the levelling, carried out in Czechoslovakia during 1964–1966, are presented in this paper. The solution is based on theoretical derivation of heat quantity received on the Earth surface from the solar radiation. The solar height angle represents the determining element in this procedure. In contrast to the studies of Holdahl the development of temperature gradients in microclime has been assessed by the approximation of one-year-set of observations of the temperature gradients. The tables for determining the coefficient of refraction correction in levelling have been derived using the empirical values. These tables allow assessment of the refraction corrections for given data and time of measurement, the average height of instrument in a levelling section, and approximative meteorological conditions. On the average, 80 % of refraction error in levelling could be minimized by introducing the above mentioned corrections.

(Author's summ.)

Techn. Univ. Dresden IGD 84/I 345

531.781.3:551.244:525.6
551.244:531.781.3

Wilmes, Herbert
Entwicklung eines horizontalen Stabextensometers mit hydrostatischer Lagerung zur Messung lokaler Gezeitendeformationen der Lithosphaere
Diss.
Dt. geod. Kommiss., R. C. – Muenchen (1983) 281. – 111 p. : div. fig. div. tab. div. lit.
Development of a horizontal rod extensometer with hydrostatic bedding to measure local tidal deformations of the lithosphere (GER)
AS 12.15.09/12.25.21/73.19
Key words: Earth's crustal movement, extensometer measurement, tidal observations

Description of an extensometer with a hydrostatically bedded length standard. Smallest variations in length may be determined with a high precision by means of a mechanical strain sensor. The transformation of the measuring movement into an appropriate voltage signal is performed by an alternating-current bridge. The calibration device permits relative and absolute calibrations. The instrumental system is well suitable for evidence and supervision of tectonically active zones.

Techn. Univ. Dresden IGD 84/I 346

551.241:528.02

Yanagisawa, Michio
Crustal movement observation data catalog and secular tilt change at the Matsuyama observatory for the period from 1949 to 1982 (JAP)
Bull. Earthq. Res. Inst. – Tokyo 58 (1983) 4. – p. 439–464 : 13 fig. 5 tab. div. lit.
AS 12.15.09/73.19/49.31
Key words: Earth's crust movement, physical geodesy, tiltmeter

The purpose of the present study is to clarify contents of the data which have been obtained at the Matsuyama Crustal Movement Observatory since September 1,

1949. Routine observations have been carried out by using water-tube tiltmeters (25 m long, two components), fused-silica extensometers (25 m long, two components and 6 m long, one component) and Ishimoto's horizontal pendulums. The observations by the extensometers and Ishimoto's pendulums were stopped in January, 1971. Except for the period from February, 1971 to January, 1972, the observations by the water-tube tiltmeters have been continued until today.

Almost all the data of the these observations still remain unpublished. In this paper, the data compilation method and the file system are considered for contributing to the future scientific usages of this data.

Nagasawa, National Correspondent, Japan IGD 84/I 347

551.244:519.2

Zakarjavichjus, A. B. / Ruzgene, B. A.
Nekotorye voprosy verojatnostno-statisticheskogo analiza vertikal'nykh dvizhenijj zemnojj poverkhnosti
Tr. po geod. Vil'njus. inzh.-stroit. in-t. — Vil'njus (1980) 10. — p. 42–54 : 12 lit.
Some problems of the statistical analysis of vertical crustal movements (RUS)
AS 12.15.09/12.15.37
Key words: Earth's crust movement, mathematical statistics

The stochastic approach is used for reduction of measurements of the vertical crustal movements. The measured values of rates of vertical movements are interpreted as a discrete realization of a stochastic function. The function contains two components. The first one is represented by actual values of regular movements due to dominant factors on the territory studied and the second one consists of purely random quantities due to measuring errors and to some random factors. The hypothesis of existence of a systematic trend of vertical crustal movements is tested by canonical expansion of stochastic functions.

Dobrokhotov, National Correspondent, USSR IGD 84/I 348

528.247

Zazuljak, P. M. et al.
Pole dispersijj anomalijj sily tjazhesti i geologicheskie struktury materikov
Geod., Kartogr. i Aehrofotos"jomka. — L'vov (1982) 36. — p. 43–49 : 6 lit.
The field of gravity anomaly variances and the geological structures of continents (RUS)
AS 12.15.09
Key words: Gravity anomaly, gravity field

The gravity field of the Earth is devided into regions according to the values of the gravity anomaly variances. Based on gravity anomalies for 1,654 equal-area blocks, gravity anomaly variances have been estimated. The totality of these estimates is considered as the variance field. For the global gravity anomaly field, 6 regions stationary according to the variance have been located. For some regions of the Earth anomalously high variances (more than 1,400 $mGal^2$) occur though for 84 % of the Earth's surface the values of variances do not exceed 400 $mGal^2$.

Dobrokhotov, National Correspondent, USSR IGD 84/I 349

528.27-187.4

...
Report on high precision gravimetry — Volume II (ENG)
IAG, Special Study Group 3.37 (ed. by Groten, Erwin et al.)
Nachr. Karten- u. Vermess.-Wesen, R. II. — Frankfurt/M. (1983) 41. — 103 p. : 4 fig. 4 tab. div. lit.
AS 12.15.09
Key words: Gravity measurement, physical geodesy

This report describes in condensed form the major topics of high precision gravity measurements. Based on highly specialized gravimetric experience and the work carried out by the various groups pertaining to IAG-Special Study Group 3.37 "Special Techniques in Gravimetry", seven experts compiled the main results of research done in the last six years. The report should, just like the first one, serve as a reference and guiding line for all persons concerned with high precision gravity work.

(Author's summ.)

Techn. Univ. Dresden IGD 84/I 350

AIDOS Systematics **12.15.13** **Group I Geodesy**

Geodetic astronomy/astrometry

Fundamentals of geodetic astronomy
Astronomic determinations of latitude, longitude and azimuth in geodetic networks
Time determination, time service
Astrometry
– Materials, methods, devices for geodetic-astronomic measurement
– Star catalogues, star co-ordinates
– Earth's rotation, polar motion
Observatories, instruments and methods of astronomic observations
– Prerequisites to carry out geodetic-astronomic measurements
– Instruments and devices for observation
– Radio astronomy

Gruppe I Höhere Geodäsie

Geodätische Astronomie/Astrometrie

Grundlagen der geodätischen Astronomie
Astronomische Breiten-, Längen und Azimutbestimmungen in geodätischen Netzen
Zeitbestimmung, Zeitdienst
Astrometrie
– Materialien, Verfahren, Geräte für geodätisch-astronomische Messungen
– Sternkataloge, Sternkoordinaten
– Erdrotation, Polbewegung
Observatorien, Instrumente und Methoden astronomischer Beobachtungen
– Voraussetzungen für die Ausführung geodätisch-astronomischer Messungen
– Beobachtungsinstrumente und -geräte
– Radioastronomie

523.164:528.28

Alekseev, V. A. / Nikonov, V. N. / Troickijj, V. S.
Radioastrometrija: sostojanie i perspektvy
Uspekhn. fiz. nauk. – (1983) 2. – p. 359–360 : 8 lit.
Radioastrometry: state and prospects (RUS)
AS 12.15.13/12.15.17
Key words: Radioastronomy

The increasing role of the methods using radiointerferometry for solution of fundamental scientific and practical problems of astrometry, geodesy, and geophysics is pointed out. These problems include: establishment of a non-rotating celestial co-ordinate system with stability of axes of 0.001″ during decades and 0.01″ during centuries; measurement of lengths of interferometric bases and geodynamic parameters with accuracy of 2 to 3 cm for lengths up to 6,000 km and 5 to 7 cm for lengths up to 8,000 km, measurement of polar co-ordinates with accuracy of 5 cm and the universal time (UT1) for diurnal observational period with accuracy of

0.002". In the USSR the project Polygam for developing a specialized satellite radiointerferometric complex for synchronization of the time scales of radiointerferometers with subsequent transfer of measuring data to processing centre has been developed.

Dobrokhotov, National Correspondent, USSR IGD 84/I 351

528.28(494):629.783

Beutler, G.

Loesung von Parameterbestimmungsproblemen in Himmelsmechanik und Satellitengeodaesie mit modernen Hilfsmitteln

Astronom.-geod. Arb. in der Schweiz hrsg. v. d. Schweiz. Geodaet. Kommiss. (Organ d. Schweiz. Naturforsch. Ges.). — Bern (1982) 34. — 257 p. : 13 fig. 15 tab. 40 lit.

Solution of problems concerning parameter determination in celestial mechanics and satellite geodesy by modern auxiliary means (GER)

AS 12.15.13/12.15.17

Key words: Geodetic astronomy, satellite geodesy

— Mathematical fundamentals and technical auxiliary means

— Problem of parameter determination of classical celestial mechanics

— Problems of parameter determination in satellite geodesy

Techn. Univ. Dresden IGD 84/I 352

522.43

Buerki, B. / Kahle, H.-G. / Schmid, H. H.

Das neue Zenitkamera-Messsystem am Institut fuer Geodaesie und Photogrammetrie der ETH Zuerich:

Ein Beitrag zur astronomisch/geodaetischen Messtechnik fuer Lotabweichungsmessungen im Gebiet der Ivrea-Zone

Vermess. Photogramm. Kulturtechnik. — Zuerich 81 (1983) 10. — p. 349–354 : 6 fig. 21 lit.

The new measuring system of the zenith camera at the Institute for Geodesy and Photogrammetry of the EHT (Swiss University of Technology) of Zurich (GER)

AS 12.15.13/47.51/12.15.09

Key words: Geodetic astronomy, vertical deflection, zenith camera

A photograph of a part of the sky in vertical direction (zenith of the station) permits the definition of the direction of the local gravity vector (vertical line) by connection to the stellar system. Due to topographic pertubation masses ect. in the surrounding of the station, these directions show deflections from the normal line of the Earth's ellipsoid — vertical deflections.

Vertical deflections combined with measured values of gravity vectors (gravity anomalies) permit significant improvements of the interpretation of the Earth's crust and the upper mantle under the Alps.

(Author's summ., abbr.)

Techn. Univ. Dresden IGD 84/I 353

621.375.8:529.786

Busca, Giovanni

The atomic hydrogen maser type EFOS 1 (ENG)

Veroeff. Bayer. Kommiss. internat. Erdmess. — Muenchen (1982). — astron. geod. Arb. 42. — p. 44–50 : 3 fig. 2 lit.

AS 12.15.13/12.15.16

Key words: Hydrogen maser, time standard, VLBI

The hydrogen maser is the most effective atomic frequency standard for averaging times between 10 and 1000 sec., i.e. in the range of interest to very long baseline interferometry. It consists of a physical package, a receiver and monitoring and control system.

(Author's summ.)

Techn. Univ. Dresden IGD 84/I 354

532.164:681.787
528.28:629.783

Campbell, James / Nottarp, Klemens
Entwicklung und Aufbau eines GPS-Interferometers
Veroeff. Bayer. Kommiss. Internat. Erdmess. : astron.-geod. Arb. — Muenchen (1981) 41. — p. 109—120 : 5 fig. 2 tab. 15 lit.
Development and set-up of a GPS interferometer (GER)
AS 12.15.13/12.15.17
Key words: Interferometer, satellite geodesy, satellite navigation

A new type of satellite navigation system, designated NAVSTAR-GPS (Global positioning system) and which is being set-up at present, offers promising possibilities for highly accurate relative point determination in the geodetic field. Various procedures are suggested, some working with, some without information on the codes GPS signals. The structure of the signals is such that both phase measurements with their time changes as well as transit time measurements are possible. Phase stability is maintained by atom frequency standards on board the satellites.

A GPS interferometer is being developed and the prerequisites created for conducting a trial experiment. The test configurations are described. Evaluation of the video tapes is done on the MK II correlator of the Max Planck Institute for Radio Astronomy in Bonn

Schoen, National Correspondent, Federal Republic of Germany IGD 84/I 355

522.4:528.28

Darakchiev, Cvetan
Vljanie na eleptichnostta i neravenstvoto na capfite na astronomicheskija universalen instrument vahrkhu nabljudenijata za azimut no direktnija metod
Vissha Geod. — Sofija (1983) 9. — p. 88—100 : 3 fig. 1 tab.
Influence of ellipticity and size-difference of the universal astronomical instrument journals on the azimuthal observations by direct method (BUL)
AS 12.15.13/47.51
Key words: Astronomical instrument, geodetic astronomy

The effect of the ellipticity and size-difference of the horizontal axis journals of universal astronomical instrument, on the measurement for determination of the azimuth by the direct method — i.e. by observing the stellar passages through the vertical of the terrestrial object, is deduced.

The formulae-corrections obtained are necessary both for correcting the current azimuth observations and for the deduction of ellipticity and size-difference of the universal instrument journals by means of astronomical observations.

(Author's summ.)

Georgiev, National Correspondent of Bulgaria IGD 84/I 356

522.21:523.164"71"

Nottarp, Klemens
Entwicklungstendenzen bei Eingangsstufen fuer Radioteleskope
Veroeff. Bayer. Kommiss. Internat. Erdmess. : astron.-geod. Arb. — Muenchen (1981) 41. — p. 106—108 : 2 fig. 6 lit.
Development trends with regard to input stages for radio telescopes (GER)
AS 12.15.13/47.51
Key words: Radio telescope

An earlier paper mentioned the fact that the development of input amplifiers for radio telescopes is progressing well and that for this reason the decision on the receiver concept for the Wettzell radio telescope should be postponed as long as possible. In order to explain this procedure, the present state in the technical development of highly sensitive input amplifiers in the frequency range 0.6 GHz to 15 GHz is described in some detail.

Schoen, National Correspondent, Federal Republic of Germany IGD 84/I 357

528.28:525.731

Rodichkin, N. N.
Opyt opredelenija astronomicheskojj refrakcii po materialam nabljudenijj opornykh zvjozd
Geod. i fotogramm. – Rostov n/D. (1981). – p. 37–45 : 8 lit.
Determination of the astronomic refraction from fixed star observation data (RUS)
AS 12.15.13
Key words: Geodetic astronomy, refraction

Dobrokhotov, National Correspondent, USSR IGD 84/I 358

528.28

Sanchez, Rafael N. / Rock Santerre
Analyse de certaines determinations astronomiques appliquees au controle d'orientation
Le Geometre Canadien. – 37 (1983) 2. – p. 71–78
Analysis of some astronomical determinations applied to direction control (FRE)
AS 12.15.13
Key words: Astronomical azimuth, Black's method

Astronomic latitude, longitude and, particularly azimuth, can be computed by Black's method from observations of the horizontal angles between a reference object and a number of stars, and the times of these observations. This article compares the accuracies of the two azimuths and draws some interesting conclusions. The article also explains the significance of azimuths computed by applying latitude and longitude taken from a map (as compared to geodetic latitude and longitude) for the astronomic observations of a number of stars. Finally, this article provides a good explanation of the role of the geodetic longitude at the initial Black's station when establishing azimuth control.

Mainville, National Correspondent, Canada IGD 84/I 359

522.614.089.6:528.28

Zernecke, Rudolf / Kaniuth, Klaus
Untersuchung der Restfehler von BMK 75-Sternaufnahmen auf Verzeichnung
Veroeff. Bayer. Kommiss. Internat. Erdmess. : astron.-geod. Arb. –
Muenchen (1981) 41. – p. 37–46 : 5 fig. 1 tab. 10 lit.
Investigation into the residual errors due to distortion in BMK 75 stellar photographs (GER)
AS 12.15.13/47.51
Key words: Calibration, stellar photograph

In the first part of this paper the results of determination of the primary comparator errors of a Wild STK 1 and a Zeiss PK 1 are given; these were obtained with the aid of a calibrated test grid. Subsequently the computation is described of the lens distortion of a BMK 75 from the residual errors in stellar photographs by means of interpolation with least squares as well as interpolation by adjustment with multiquadratic functions. Distortion attains a maximum of 3 μ and is

represented planimetrically due to its irregularity. In this particular application, the multiquadratic method proved more satisfactory than interpolation according to least squares.
Schoen, National Correspondent, Federal Republic of Germany IGD 84/I 360

AIDOS Systematics **12.15.17** **Group I Geodesy**

Space geodesy

Techniques of satellite geodesy for determining the figure of the Earth
Solving geodetic problems with the aid of artificial earth satellites
Geometric and dynamic methods of satellite geodesy
Satellite networks
Satellite observation, methods and means
Observation of the Moon and of distance space objects for geodetic purposes
Laser ranging to the Moon
Very-long-base interferometry

Gruppe I Höhere Geodäsie

Kosmische Geodäsie

Verfahren der Satellitengeodäsie zur Bestimmung der Erdfigur
Lösung geodätischer Aufgaben mit Hilfe künstlicher Erdsatelliten
Geometrische und dynamische Methoden der Satellitengeodäsie, Satellitennetze
Satellitenbeobachtung, Methoden und Mittel
Beobachtung des Mondes und von entfernten kosmischen Objekten für geodätische Zwecke
Laserortung des Mondes
Very-long-base-interferometry

528.22:629.783

Borisov, Eh. A.
O vozmushhenijakh ehlementov orbity blizkikh iskusstvennykh sputnikov
Geod., Kartogr. i Aehrofotos"jomka. – L'vov (1982) 36. – p. 7–10
On orbital element perturbations of near satellites (RUS)
AS 12.15.17/12.15.09
Key words: Orbital element, physical geodesy, satellite geodesy

When using the systems for compensation of non-gravitational forces, the geodetic and non-geodetic satellites with heights of up to 200 km acquire great importance for solution of the problems of satellite geodesy. But in this case the orbital elements are considerably perturbed. A formula for the perturbation function under condition of secular variations of angular elements has been derived. The first-order-perturbations are obtained by integration of the elliptical motion equations. For Levi-Chevitte-parameter introduced into the orbital element system instead of the eccentricity, the perturbations due to the 3d- and 4th-degree zonal harmonics of the geopotential are obtained.
Dobrokhotov, National Correspondent, USSR IGD 84/I 361

528.021.78:629.783
528.517.8.089.6

Brandl, Nikolaus
Messaufbau zur internen Laufzeitmessung des Nd : YAG-Laser-Entfernungsmesssystems der Station Wettzell
Veroeff. Bayer. Kommiss. Internat. Erdmess. : astron.-geod. Arb. – Muenchen (1981) 41. – p. 49–52 : 2 fig. 3 lit.
Measuring arrangement for internal transit time measurements with the Nd : YAG laser distance measuring system of the Wettzell station (GER)
AS 12.15.17/12.15.33/73.07
Key words: Laser distance measurement

In 1980, as part of the work done with a view to increase accuracy, the position of the zero point was determined on the Nd : YAG laser distance measuring system. A description is given here of the measuring arrangements of internal transit time measuring. One procedure is based on the addition of sections; regarding the other, very short internal transit times in the system are determined through a retro prism.

Schoen, National Correspondent, Federal Republic of Germany IGD 84/I 362

528.22:629.783

Chapanov, Javor / Georgiev, Ivan
Opredeljane na promezhdutahchna orbita za poljarni ISZ
Vissha Geod. — Sofija (1983) 9. — p. 19—24
Determination of intermediate orbit of polar artificial Earth satellites
(BUL)
AS 12.15.17/12.15.09
Key words: Physical geodesy, satellite geodesy

Aiming to obtain better convergent power series of variable ω by the regularized time τ, in the intermediate motion of artificial Earth satellites according to the generalized problem of two fixed centres, power series of right ascension of node Ω by regularized time τ are used, which series are characterized by a speed convergence. The reccurent formulae deduced enable one to determine the coefficients of any desired order in the power series development of Ω. The method described is efficient when determining the intermediate orbit of a satellite the inclination of which is close to 90°.

(Author's summ.)

Georgiev, National Correspondent of Bulgaria IGD 84/I 363

528.22:629.783

Daskalova, Mara
O konstantakh c i σ **potenciala promezhutochnogo gravitacionnogo polja sozdannogo dvumja centrami pritjazhenija**
Vissha Geod. — Sofija (1983) 9. — p. 25—30 : 2 tab.
On the constants c **and** σ **in the potential of intermediate gravity field generated by two attracting centres** (BUL)
AS 12.15.17/12.15.09
Key words: Gravity field, physical geodesy, satellite geodesy

Two approaches of applying of an integral criterion are suggested for the determination of the constants c and σ in the expression for the intermediate gravity field generated by two attracting centres. The Earth potential is given by zonal spherical harmonics of order up to $n = 4$, while the potential of the intermediate field is represented in two different forms: by means of spherical harmonics of order up to $n = 4$, too, and by a closed formula. The method suggested could be applied to the case $n > 4$, too, when this is desirable.

(Author's summ.)

Georgiev, National Correspondent of Bulgaria IGD 84/I 364

528.22:629.783
521.93

Khadzhijjski, A.
Ob optimal'nom vyborc orbity USZ s cel'ju izuchenija dvizhenija poljusa
Vissha Geod. — Sofija (1983) 9. — p. 41—45 : 1 fig. 2 tab.
On the optimum choice of artificial Earth satellite orbit for studying the polar motion (BUL)
AS 12.15.17/12.15.09/12.15.13
Key words: Geodetic astronomy, polar motion, satellite geodesy

The connection between the accuracy of determination of the Earth pole position and the geometry of the satellite orbit, located by means of laser observations, is examined. According to the model investigations performed, the following conclusions have been drawn: the main semiaxis and the inclination of the artificial Earth satellite orbit exert largest influence on the accuracy; the high polar satellites are most convenient for studying the polar motion. The use of such satellites provides for an accuracy of the order of 0.0033″ for a diurnal interval.

(Author's summ.)

Georgiev, National Correspondent of Bulgaria IGD 84/I 365

528.22:629.783

Koceva, Veneta / Georgiev, Nikola
Stepennye rjady vozmuzhhenijj v promezhutoehnom dvizhenii ISZ, vyzvannykh pritjazheniem atmosfery
Vissha Geod. – Sofija (1983) 9. – p. 12–18
Power series for perturbations in intermediate motion of artificial Earth's satellites due to atmospheric attraction (BUL)
AS 12.15.17/12.15.09
Key words: Physical geodesy, satellite geodesy

Operational formulae are deduced for the determination of perturbations due to the potential of the joint earth-atmospheric attraction in the differential equations of the intermediate motion for short time intervals. The power series obtained by regularized time yield separately the perturbations from zonal harmonics and the perturbations from the tesseral and sectorial harmonics of the potential. The method described of taking into consideration the atmospheric attraction enables one to use the same coefficient both for the zonal and for the tesseral harmonics of the general Earth's and atmospheric attraction; this method makes it possible to solve, by a similar manner, the problem of determining the perturbations from all the harmonics of the summary potential, too.

(Author's summ.)

Georgiev, National Correspondent of Bulgaria IGD 84/I 366

528.28.021.78:629.783
521.26:629.783

Mueller, Horst / Reigber, Christoph / Wende, Werner
Geozentrische Koordinaten der Stationen 7835 und 7120, abgeleitet aus Lageos-Beobachtungen
Veroeff. Bayer. Kommiss. Internat. Erdmess. : astron.-geod. Arb. –
Muenchen (1981) 41. – p. 146–153 : 4 fig. 2 tab. 9 lit.
Geocentric co-ordinates of the stations 7835 and 7120, derived from Lageos observations (GER)
AS 12.15.17/12.15.09
Key words: Geocentric co-ordinates, satellite geodesy

This paper deals with the determination of the geocentric co-ordinates of the satellite observation stations 7835 (Grasse, France) and 7120 (Maui, Hawaii), constituting part of the work on polar co-ordinate determination from observation data of the MERIT project. The co-ordinates are determined dynamically relative to the reference system which is ascertained with the aid of the co-ordinate solution of the University of Texas and on the basis of the provisional GRIM3 gravitation model.

Schoen, National Correspondent, Federal Republic of Germany IGD 84/I 367

528.22:629.783

Nasretdinov, K. K.
Ocenka tochnosti polozhenija i skorosti sputnika
Geod. i Kartogr. – Moskva (1982) 11. – p. 15–18 : 5 lit.
Estimation of satellite positioning and velocity measurement accuracy (RUS)
AS 12.15.17
Key words: Satellite geodesy, satellite positioning

For the orbital method of development of satellite geodetic nets the formulas for estimation of the position and the velocity of a satellite as well as their accuracy at the time moment given both in the three-dimensional rectangular co-ordinate system and in the orbital one have been derived. The said formulas may be used for solution of applied problems of satellite geodesy.

Dobrokhotov, National Correspondent, USSR IGD 84/I 368

528.516:621.396.962

Nottarp, Klemens / Schlueter, Wolfgang
Das geplante TRANET 2-Doppleremfangssystem fuer die Satellitenbeobachtungsstation Wettzell
Veroeff. Bayer. Kommiss. Internat. Erdmess. : astron.-geod. Arb. – Muenchen (1981) 41. – p. 91–97 : 5 fig. 1 tab. 8 lit.
The TRANET 2 Doppler receiver system planned for the Wettzell satellite observation station (GER)
AS 12.15.17/12.15.09
Key words: Doppler observation, satellite geodesy

The decision between a mobile and a stationary Doppler system for the Wettzell station fell on the stationary system TRANET 2 developed by the Physical Science Laboratory, University of New Mexico, Las Cruces, U.S.A. The system carries out fully automatic Doppler observations on both frequency pairs (150/400 MHz and 162/324 MHz), controls the receiver functions independently and checks the data. The main advantages of the system are its very high reliability and easy maintenance. The system was conceived for long-term measurements; the building block construction together with a micro-computer allows for continual adaptation to future needs.

The set-up of the TRANET 2 Doppler measuring systems and the concept of the antenna for the Wettzell station are described.

Schoen, National Correspondent, Federal Republic of Germany IGD 84/I 369

528.34.088:629.783

Pleshakov, I. Ja. / Nikolaev, A. S. / Ieshin, E. G.
Tochnost' opredelenija vzaimnogo polozhenija punktov po rezul'tatam nabljudenijj ISZ
Geod. i Kartogr. – Moskva (1982) 10. – p. 14–18
The accuracy of determination of relative station position from satellite observation data (RUS)
AS 12.15.17/12.15.37
Key words: Geodetic network, mathematical geodesy, positioning, satellite geodesy

R.M.S. errors of position of the stations of a satellite geodetic net established from the simultaneously measured components of the vector "station-satellite" are numerically estimated. The computations of these R.M.S. errors are based on the mean integral estimates of geometrical and statistical net parameters. The increasing accuracy of orientation of chords connecting the stations to be determined with increasing of satellite height has been estimated.

Dobrokhotov, National Correspondent, USSR IGD 84/I 370

528.28:629.783

Seeber, G.
Die Rolle des NAVSTAR Global Positioning Systems fuer die Loesung geodaetischer Aufgaben
Z. Vermess.-Wesen. — Stuttgart 109 (1984) 1. — p. 1—11 : 5 fig. 3 tab. div. lit.
The role of the NAVSTAR Global Positioning System for the solution of geodetic tasks (GER)
AS 12.15.17
Key words: Global positioning system, navigation system with time and ranging

This survey paper describes the concept, the status and the instrumental developments connected with the Global Positioning System. The different observation techniques are discussed together with the achievable accuracy. Some current and future applications in geodesy are outlined.

(Author's summ.)

Techn. Univ. Dresden IGD 84/I 371

528.22:629.783

Stepin, Ju. D.
Ob uchjote vlijanija geopotenciala pri vysokotochnom chislennom integrirovanii orbit ISZ
Izvest. vyssh. uchebn. Zaved., Geod. i Aehrofotos"jomka. — Moskva (1982) 4. — p. 93—97 : 6 lit.
On accounting for the effect of the geopotential in high-precision numerical integration of satellite orbits (RUS)
AS 12.15.17/12.15.09
Key words: Geopotential, satellite geodesy

Dobrokhotov, National Correspondent, USSR IGD 84/I 372

528.23:629.783

Stojanov, Ljubomir
Sumirane na redove pri rekurentni zavisimosti v zadachite za dvizhenie na ISZ
Vissha Geod. — Sofija (1983) 9. — p. 31—40
Series summation under recurrent dependences in the problems for artificial Earth satellite motion (BUL)
AS 12.15.17/12.15.05
Key words: Mathematical geodesy, satellite geodesy

Some recurrent processes of calculation the final sums, the coefficients or the special functions of which meet the requirements of the homogeneous recurrent formulae, are described. The problems of an elliptical motion, of perturbations computation due to the zonal and angular parts of the potential, of determination of the perturbing acceleration in the intermediate motion etc. are discussed, too. The recurrent processes mentioned could be applied both in analytical and in numerical methods when integrating the motion equations of the artificial Earth satellites.

(Author's summ.)

Georgiev, National Correspondent of Bulgaria IGD 84/I 373

528.22:629.783
521.3:528.711.18

Stolbov, I. A.
Vlijanie orbital'nojj aberracii na polozhenie tochek kosmicheskikh snimkov
Izvest. vyssh. uchebn. Zaved., Geodez. i Aehrofotos"jomka. — Moskva (1982) (1982) 5. — p. 98—101 : 3 lit.
The effect of orbital aberration on position of points of space images (RUS)
AS 12.15.17/12.15.13/12.29.13
Key words: Orbital aberration, physical geodesy, satellite geodesy, space image

Displacement of points on space images due to the orbital aberration is discussed. The formulas allowing for such a displacement for points of vertical and oblique images are derived. The possibility of the exact accounting for the said effect is substantiated.
Dobrokhotov, National Correspondent, USSR IGD 84/I 374

528.33:621.396.962:629.783

Swiatek, Krysztof
Anwendung von Doppler-Satellitenmessungen zur Genauigkeitsverbesserung geodaetischer Netze
Z. Vermess.-Wesen. — Stuttgart 109 (1984) 2. — p. 65—74 : 8 tab. div. lit.
The application of Doppler satellite measurements for accuracy improvements of geodetic nets (GER)
AS 12.15.17/12.15.09
Key words: Doppler measurement, satellite geodesy

A combined adjustment of a Doppler satellite and a terrestrial network is presented. The numerical calculations are performed in the testnet Western Harz in three dimensions as well as in two dimensions by algebraic and geometrical elimination of heights. The results indicate a clear diminution of the maximum values of the mean coordinate errors when Doppler stations are included. (Author's summ.)
Techn. Univ. Dresden IGD 84/I 375

528.2:629.783

Tamutis, Z.
Ispol'zovanie kosmicheskikh ob"ektov v geodezii
Tr. po geod. Vil'njus. inzh.-stroit. in-t. — Vil'njus (1982) 11. — p. 84—90
The use of space objects in geodesy (RUS)
AS 12.15.17
Key words: Satellite geodesy

In a brief review the role of space object observations for study of the dimensions, the form and the outer gravity field of the Earth, for establishing the co-ordinate systems, polar motion studies, for determination of position of ground points and terrestrial space in a reference system and in a general one, for studies of geodynamical phenomena and other problems is shown. It is emphasized that in modern geodesy the study of the metric of nature at cosmic scale is becoming one of the fundamental geosciences.
Dobrokhotov, National Correspondent, USSR IGD 84/I 376

528.02:528.2:629.783

Tatevjan, S. K.
Ispol'zovanie opticheskikh nabljudenijj iskusstvennykh sputnikov Zemli dlja celejj geodezii i geodinamiki
Uspekhi fiz. nauk. — (1983) 2. — p. 357—360 : 12 lit.
The use of optical observations of satellites for geodesy and geodynamics (RUS)
AS 12.15.17/12.15.33
Key words: Satellite geodesy

The achievments in satellite geodesy using photographic and laser observation techniques are discussed. Future development of satellite geodesy is directed to the study of polar motion and Earth rotation irregularities. In the coming years new laser distance measuring equipments are expected to be put into action in some countries that will allow determination of polar position with the accuracy of 5 cm during 1 day observation period. It is recommended to perform laser observations of high satellites for determination of geodynamic parameters on the intervals of up to 2 days and to use Lunar ranging data for determination of time corrections on longer intervals.
Dobrokhotov, National Correspondent, USSR IGD 84/I 377

528.22.02:629.783

Varga, M.
Review and comparison of recent methods in space geodesy (ENG)
Period. Polytechn. : Ser. Civil Engin. — Budapest 27 (1983) 3—4. — p. 253—259 : 1 tab. 15 lit.
AS 12.15.17/12.15.33
Key words: Laser altimetry, radio interferometry, space geodesy

Studying geodynamic processes requires us to make use of new-type space-borne geodesy measurements. A terrestrial reference system (TRS) is required for describing geodynamic processes. For this we should have satisfactory knowledge of polar motion, Earth rotation and tidal forces, determined by laser, GPS and VLBI measurements. In addition, gravity and magnetic field of the Earth have to be known, modelled by making use of SST, altimetry, gradiometry and magnetometry results. Motion of the Earth-Moon system, as well as the relation between the terrestrial reference system and the inertial system can be determined by means of VLBI measurements.
Veroe-Hetenyi, National Correspondent, Hungary IGD 84/I 378

AIDOS Systematics **12.15.21** **Group I Geodesy**

Marine geodesy

Determination of the figure of the Earth below seas and oceans
Positioning on the water surface
Variations of the sea level — oceanic tides
Methods and means of measurements of marine geodesy
Submarine earthquakes and geodetic measurements

Gruppe I Höhere Geodäsie

Meeresgeodäsie

Bestimmung der Figur der Erde unter Meeren und Ozeanen
Lagebestimmung auf der Wasseroberfläche
Schwankungen des Meeresniveaus — Meeresgezeiten
Methoden und Mittel der Messungen der Meeresgeodäsie
Meeresbeben und geodätische Messungen

621.396.969.11:629.783
528.21(261.26)

Brennecke, Juergen
Zur Altimetergeoidberechnung im Rahmen des SEASAT-1-Nordsee-Projekts
Veroeff. Bayer. Kommiss. Internat. Erdmess. : astron.-geod. Arb. — Muenchen (1981) 41. — p. 131—145 : 4 fig. 3 tab. 12 lit.
Computation of the altimetric geoid within the SEASAT 1 North Sea project (GER)
AS 12.15.21/12.15.33/12.15.37
Key words: Altimetry, geoid, marine geodesy, satellite geodesy

This paper presents the first results of determination of the mean sea surface in the test area. The computation model on which the two-dimensional computations are based and the data material are described. Two further sections deal with partial aspects of altimetric geoid determination, i.e. correction of altimeter measurement due to the effect of wave height as well as the influence of short-arc orbit determination on computation of the mean sea surface.
Schoen, National Correspondent, Federal Republic of Germany IGD 84/I 379

525.622:528.28

Galkin, V. Ja. et al.
Issledovanie matematicheskojj modeli zemnykh prilivov
Obrab. i interpretacija rezul'tatov nabljudenijj. – Moskva (1981). – p. 74–84 : 8 lit.
The study of a mathematical model of Earth tides (RUS)
AS 12.15.21/12.15.13
Key words: Earth tides, Love parameter, marine geodesy

A mathematical model for determination of Love parameter k is studied. The model is obtained by using the law of conservation of motion momentum and the tidal component of Earth's rotation irregularity determined from astronomical observations. From calculations for five groups of M-waves the Love number k was estimated that is in a quite good agreement with k-value for Molodensky's Earth model with a liquid outer core and a solid interior one. Among the five versions of the mathematical model discussed the stationary and the non-stationary models with phase shift have been acknowledged as the best ones.
Dobrokhotov, National Correspondent, USSR IGD 84/I 380

528.27(26):551.241

Glebovskijj, V. Ju. / Litvinov, Eh. M.
Ocenka moshhnosti zemnojj kory Atlanticheskogo okeana po gravimetricheskim dannym
Mor. geofiz.-issled. v Arktike. – Leningrad (1981). – p. 46–51
The estimation of crustal thickness in the Atlantic from gravity data (RUS)
AS 12.15.21/12.15.09/12.15.33
Key words: Gravity measurement, marine geodesy
Dobrokhotov, National Correspondent, USSR IGD 84/I 381

528.27(26):551.241

Kosygin, V. Ju. / Pavlov, Ju. A. / Stroev, P. A.
Osnovnye cherty gravitacionnogo polja, moshhnost' i izostaticheskoe sostojanie zemnojj kory severo-zapadnogo sektora Tikhogo okeana
Tikhookean. geol. – Moskva (1982) 1. – p. 49–59 : 22 lit.
Main features of the gravity field, thickness and isostasy of the crust in the North-Western Pacific (RUS)
AS 12.15.21/12.15.09/12.15.33
Key words: Earth'crust, gravity field, marine geodesy
Dobrokhotov, National Correspondent, USSR IGD 84/I 382

528.27(26):551.462

Lukashevich, I. P. et al.
Svjaz' gravitacionnogo polja i topografii dna v rajjone Severnojj Atlantiki s dinamicheskimi processami v mantii
Geol. morejj i okeanov. Tez. dokl. 5 Vses. shk. mor. geol. T. 2. – Moskva (1982). – p. 51–53
The relation of the gravity field and the sea-bottom topography in Northern Atlantic with dynamical processes in the mantle (RUS)
AS 12.15.21/12.15.09/12.15.33
Key words: Gravity field, marine geodesy
Dobrokhotov, National Correspondent, USSR IGD 84/I 383

AIDOS Systematics **12.15.25** **Group I Geodesy**

Establishing of geodetic control networks

Network design, optimum network configuration
Geodetic-astronomic control points
Measurements and evaluations of geodetic networks

Gruppe I Höhere Geodäsie

Anlage geodätischer Festpunktnetze

Projektierung von Netzen, optimale Netzkonfiguration
Geodätisch-astronomische Fundamentalpunkte
Messungen und Auswertungen geodätischer Netze

528.3:061.1(430.1-17)

Augath, Wolfgang
Moderne Grundlagenvermessung im norddeutschen Raum. Anforderungen und Realisierung
Z. Vermess.-Wesen. – Stuttgart 109 (1984) 1. – p. 12–19 : 5 fig. 4 tab. div. lit.
Modern control survey in Northern Germany. Demands and their realization (GER)
AS 12.15.25
Key words: Control network

The possibilities of new control survey techniques are discussed from the point of view using them in horizontal, vertical and gravity control networks. It is emphasized that a strict distinction of these networks is no longer opportune and that the availability of the results has to be more flexible.

(Author's summ.)

Techn. Univ. Dresden IGD 84/I 384

528.33:528.18:519.28

Banhegyi, Istvan / Dede, Karoly
Adjustment of horizontal control point networks by partitioning (ENG)
Period. Polytechn. : Ser. Civil Engin. – Budapest 27 (1983) 3–4. –
p. 171–181 : 1 fig. 1 tab. 9 lit.
AS 12.15.25/12.15.37
Key words: Adjustment of combined networks, horizontal networks, partitioning

The matrix partitioning presented has several practical advantages. It permits to solve extended normal equation systems or to find regularities in adjusting e.g. combined networks by the indirect measurement method. Use of partitioning – as shown in the numerical example – is recommended primarily for the rapid solution of direct practical problems. With the advent of high capacity computers, adjustment of a horizontal network consisting of a few points would not justify partitioning of the coefficient matrix of normal equations. We note, however, that the rapid solution of individual problems leaves no time to engage services of computing centres and in such cases adjustment can be carried out using pocket calculators or desk computers.

Veroe-Hetenyi, National Correspondent, Hungary IGD 84/I 385

528.38:528.13

Brazis, L. B. / Skejjvalas, I. M.
K voprosu vlijanija oshibok iskhodnykh dannykh na tochnost' uravnennykh velichin
Tr. po geod. Vil'njus. inzh.-stroit. in-t. — Vil'njus (1980) 10. — p. 103—106
On the effect of initial data errors upon the accuracy of adjusted values (RUS)
AS 12.15.25/12.15.37
Key words: Adjustment, error of initial data, levelling network

The effect of initial data errors upon the adjustment accuracy of a levelling net has been studied by three different methods as follows: the net was adjusted as a free one; the net was adjusted with inclusion of initial data errors simulated by Monte Carlo method in 4 versions on the supposition of normally distributed measuring errors; the net was adjusted by the generalized least squares method with determination of corrections to the initial data. The highest accuracy was achieved by the generalized least squares method. The effect of the initial data errors is found to be negligible if they are 2 to 4 times higher than the measuring accuracy.

Dobrokhotov, National Correspondent, USSR IGD 84/I 386

528.31(666.8):621.396.962.33

Cisse, A. et al.
Die Erstellung eines uebergeordneten Festpunktnetzes in der Republik Elfenbeinkueste mit Hilfe von Dopplermessungen
Allg. Vermess.-Nachr. — Karlsruhe 91 (1984) 1. — p. 3—8 : 2 fig. 1 tab. 12 lit.
The establishment of a primary network of control points in the Republic of the Ivory Coast (GER)
AS 12.15.25/12.15.09/12.15.17
Key words: Co-ordinates determination, Doppler measurement, geodetic network

A primary network consisting of 17 points has been established in the territory of the Republic of the Ivory Coast by Doppler measurements. Within this paper the measurements and the computation procedures are described. The accuracy of the co-ordinates is estimated to be better than $\pm$ 0.5 m.

(Author's summ.)

Techn. Univ. Dresden IGD 84/I 387

528.33:528.14

Hein, Guenter W. / Landau, Herbert / Egreder, Kurt
Erste Erfahrungen zur integrierten geodaetischen Netzausgleichung
Z. Vermess.-Wesen. — Stuttgart 109 (1984) 2. — p. 75—86 : 3 fig. 2 tab. div. lit.
First experiences in integrated geodetic net adjustment (GER)
AS 12.15.25/12.15.37
Key words: Geodetic network, integrated geodesy adjustment

After discussing the state-of-the art of geodetic networks we are proposing to use in future the integrated geodesy adjustment where 3d-coordinates and the functionals of the gravity disturbing potential are estimated in *one* unified model considering *all* available (terrestrial) geodetic observations. The model and related datum considerations are shortly outlined. First practical results are shown using data of the testnet "Loreley" which are processed by the new program system "OPERA" (operational adjustment). The experiences clearly demonstrate the superiority of the method over the traditional approach where the determination of deflections of the vertical was done by physical geodesy or geodetic astronomy and the pure geometrical adjustment by surveyors.

(Author's summ.)

Techn. Univ. Dresden IGD 84/I 388

528.34/.35:528.14

Markuze, Ju. I. / Rabinovich, I. E.
Sovmestnoe uravnivanie setejj na razlichnykh montazhnykh gorizontakh po principu uchjota oshibok iskhodnykh dannykh
Tr. po geod. Vil'njus. inzh.-stroit. in-t. — Vil'njus (1982) 11. — p. 46—50
A simultaneous adjustment of the nets in different planes of assembling accounting for initial data errors (RUS)
AS 12.15.25/12.15.37
Key words: Adjustment, error of initial data, geodetic network

Based on the parametric adjustment method with account for initial data errors, it is shown that the adjustment of checking measurements carried out in the assembly planes with account for co-ordinate errors of the stations obtained by vertical projection from lower planes increases considerably the accuracy of these stations. Thereby it is possible to estimate rigorously the accuracy of any station of the three-dimensional control net of the construction with respect to the co-ordinate system adopted on the initial assembly plane. An example for the case of vertical projection of a trilateration net with checking distance measurements is given.

Dobrokhotov, National Correspondent, USSR IGD 84/I 389

(528.33+528.41):528.112

Muerle, M. / Bill, R.
Zuverlaessigkeits- und Genauigkeitsuntersuchung ebener geodaetischer Netze
Allg. Vermess.-Nachr. — Karlsruhe 91 (1984) 2. — p. 45—62 : 9 fig. 1 tab. 19 lit.
Reliability and accuracy investigation of plane geodetic nets (GER)
AS 12.15.25/12.15.37/12.23.05
Key words: Geodetic network, precision and reliability of plane geodetic network

The quality of a geodetic network is described by measures of precision and reliability. These measures depend on how to define the mathematical model. The influence of model enlargement is analysed in theory and is shown in a small example. For different kinds of plane geodetic networks the measures of precision and reliability are computed and compared.

(Author's summ.)

Techn. Univ. Dresden IGD 84/I 390

528.338.004.5:629.783

Nottarp, Klemens
Konzept eines Systems zur Ueberwachung von geodaetischen Punktfeldern
Veroeff. Bayer. Kommiss. Internat. Erdmess. : astron.-geod. Arb. — Muenchen (1981) 41. — p. 121—123 : 1 fig.
Concept of a system for monitoring geodetic control nets (GER)
AS 12.15.25/12.15.17
Key words: Geodetic control net, satellite geodesy

The concept of a less expensive satellite-assisted system for the continous monitoring of densely covered nets is presented. For this reason the utilisation of television satellites in geostationary orbits, terrestrial television transmitters and mass-produced components for the measuring stations is being considered. Determination of interactive planimetric changes of the individual measuring points outside the control net is based on a phase difference measurement using measuring signals received and co-ordinated from a geostationary satellite and several terrestrial stations. The measuring principle is described in detail.

Schoen, National Correspondent, Federal Republic of Germany IGD 84/I 391

528.271(44+100)

Ogier, M.
Integration du reseau gravimetrique francais RGF 83 dans le reseau international IGSN 71
Bull. Inform. – Toulouse (1983) 53. – p. 142–149 : 5 fig. 4 tab.
Integration of the French Gravity Net RGF 83 into the International Net IGSN 71 (FRE)
AS 12.15.25/12.15.09
Key words: France, gravity network, International Gravity Net

Measurements were performed between 1980–1983, during the establishment of the new French Gravity Net, on all the French IGSN 71 stations still available and some European ones.

The results clearly show:

1) a linear correlation between the difference ($g_{CNR} - g_{RGF\,83}$) and the value of the gravity field
2) the lack of correlation between the difference ($g_{IGSN\,71} - g_{RGF\,83}$) and the value of the gravity field.

These observations cause us to ask about the validity of the periodic variation of ($g_{absolute} - g_{IGSN\,71}$) versus the gravity field as shown by Canizzo, L. et al. (1978).

In addition we have not found the $+50$ μGal systematic difference between absolute and IGSN 71 values, our data showing a difference of -71 μGal for Toulouse.

(Author's summ.)

Techn. Univ. Dresden IGD 84/I 392

528.3:528(091)

Rinner, Karl
Entwicklungstendenzen in der Landesvermessung
Acta Geodaet., Geophys. et Montan. – Budapest 18 (1983) 1–2. – p. 15–27
Trends of development in national surveys (GER)
AS 12.15.25
Key words: Control points, digital models, geodetic observatories, history of geodesy, national survey, navigation satellite

National survey involves the creation of the basis of geodetic operations required in national, regional, and continental areas. It is observed that these fundamentals are not fixed definitions but depend on requirements and theories and on available technologies. From the present point of view these are systems of terrestrial control points and navigation satellites as well as maps and digital models for conveyance of land information and geodetic observatories where the connection to higher systems is exercised.

In the following the historical development of geodesy is considered first. Then the nature and accuracy of measured variables, models, and procedures as well as education are discussed. A view to the future shows the activities in geodesy in developing countries as well as in countries with geodetic tradition.

Veroe-Hetenyi, National Correspondent, Hungary IGD 84/I 393

528.331.1:528.13

Sarkoezy, Ferenc
Optimierungsprobleme bei ausgedehnten Netzen
Period. Polytechn. : Ser. Civil Engin. — Budapest 27 (1983) 3—4. — p. 113—124 : 5 fig. 3 lit.
Optimization problems in extended networks (GER)
AS 12.15.25/12.15.37
Key words: National network, network optimization, network stiffening, optimum network

National network dimensions inhibit the application of first- and second-order design by use of existing algorithms. Objective functions for three problems facilitating the application of mathematical programming methods in national networks will be presented.

The first problem suggests a first-order design involving minimum signal building costs.

The second problem suggests a rather expedient network-stiffening technology for the Doppler translocation method.

The third problem deduces objective functions for the optimum structure of the national network.

Veroe-Hetenyi, National Correspondent, Hungary IGD 84/I 394

528.331(666.8):629.783

Schoedlbauer, Albert
Einrichtung eines geodaetischen Festpunktfeldes erster Ordnung in der Republik Elfenbeinkueste mit Hilfe des Navy Navigation Satellite Systems
Z. Vermess.-Wesen. — Stuttgart 108 (1983) 7. — p. 309—313 : 5 lit.
The establishment of the first-order control point net of the Republic of the Ivory Coast by means of the Navy Navigation Satellite System (GER)
AS 12.15.25/12.15.17
Key words: Doppler control network

Techn. Univ. Dresden IGD 84/I 395

528.38.088.3

Torim, A. A.
Vlijanie dliny vizirnogo lucha na tochnost' nivelirovanija
Tr. po geod. Vil'njus. inzh.-stroit. in-t. — Vil'njus (1980) 10. — p. 22—25
The effect of the sight lenght on levelling accuracy (RUS)
AS 12.15.25/12.15.33
Key words: Accuracy, length of sight line

Dobrokhotov, National Correspondent, USSR IGD 84/I 396

528.381(474.2)

Torim, A. A.
Analiz rezul'tatov nivelirovanija I klassa Instituta astrofiziki i fiziki atmosfery AN EhSSR
Tr. po geod. Vil'njus. inzh.-stroit. in-t. — Vil'njus (1980) 10. — p. 11—15
An analysis of the first-order levelling data in the Institute of Astrophysics and Physics of Atmosphere of the Academy of Sciences of Estonian SSR (RUS)
AS 12.15.25
Key words: Levelling

Dobrokhotov, National Correspondent, USSR IGD 84/I 397

AIDOS Systematics **12.15.33** **Group I Geodesy**

Geodetic measuring methods

Methods and techniques for geodetic measurements of distances, angles, heights
Gravity measurements
Assessment of the precision of geodetic measurements

Gruppe I Höhere Geodäsie

Geodätische Meßverfahren

Verfahren und Technologien für geodätische Messungen von Strecken, Winkeln, Höhen
Schweremessungen
Beurteilung der Genauigkeit geodätischer Messungen

528.563

Csapo, Geza
Geodeziai tipusu LaCoste-Romberg gravimeter leolvasoberendezesenek vizsgalata
Geod. es Kartogr. – Budapest 35 (1983) 5. – p. 336–340 : 2 fig. 1 tab. 8 lit.
Investigation of the reading device of a geodetic LaCoste-Romberg gravimeter (HUN)
AS 12.15.33/73.19
Key words: LaCoste-Romberg gravimeter, periodical errors, reading device

In recent years the number of high precision gravimetric measurements has been increasing all over the world. For such measurements mainly LaCoste-Romberg gravimeters are used. The required high precision demands detailed examinations of the gravimeters which are not necessary in the course of routine mineral exploration.
The paper deals with periodical errors due to imperfections of the transmissions of the reading dial. Examples are given for errors caused by imperfections of the reading device.

Veroe-Hetenyi, National Correspondent, Hungary IGD 84/I 398

331.875(528.02+528.5)

Deumlich, Fritz
Zur Mechanisierung und Automatisierung geodaetischer Instrumente und Verfahren
Vermessungstechnik. – Berlin 32 (1984) 2. – p. 41–42 : 7 lit.
Mechanization and automation of geodetic instruments and techniques (GER)
AS 12.15.33/73.07
Key words: Automation, geodetic instrument, geodetic technique, mechanization

Aim and sense of mechanization and automation in geodesy. Most important effects of recent time: EDP, electronic distance measurement, Earth's satellites, microprocessors. Developing trends in national surveying above all, by satellite geodesy and large-scale surveying, mostly by electronic tachymeters.
(Author's summ.)

Techn. Univ. Dresden IGD 84/I 399

528.33:528.14

Grimhardt, H. / Schnaedelbach, Klaus
Rigorous adjustment of horizontal geodetic networks with given tolerance intervals for the observations (ENG)
presented to: IAG General Assembly, Hamburg, 15–27 August, 1983. – p. 1–10 : 2 fig. 8 lit.
AS 12.15.33/12.15.37
Key words: Adjustment, geodetic network

In 1975 and 1976, Heindl/Reinhart presented a method for the adjustment of horizontal geodetic networks which is based on an exact estimation of minimal maximum errors when tolerance intervals are given for the observations (Minimax-method). Up to now, only small networks with simple structure can be adjusted rigorously, while large networks with arbitrary configration have to be determined approximately.

In the presented paper, a method is derived which in the sense of the above mentioned principle, allows a rigorous adjustment of horizontal networks, regardeless of their structures or dimension. Basis of the new method is the theory of non-linear parametric optimization.

The results are tolerance polyhedrons and most probable values for the positions of the new points.

(Author's summ.)

Techn. Univ. Dresden IGD 84/I 400

528.021.7:528.517

Ivanov, Ivan Cvetanov
Tochnost na dahlzhinite izmervanija s deviatorna elektrooptichna sistema
Geodez., kartogr., zemeustr. — Sofija (1983) 1. — p. 16—18 : 4 fig.
Accuracy of distance measurements by a deviator electro-optical system
(BUL)
AS 12.15.33/73.07
Key words: Electro-optical distance measurement, electro-optical rangefinder

A theoretical study on accuracy of measuring the distances by means of electro-optical rangefinders, when applying the optical deviator system suggested by the author, instead of usual reflection tripel-prisms, is performed. It is maintained that the distances obtained as a difference of two measurements are more accurate than the directly measured distances.

(Author's summ.)

Techn. Univ. Dresden IGD 84/I 401

528.02.061.2

Jakovlev, V. A.
Povyshenie tochnosti uchjota meteorologicheskikh faktorov pri opticheskojj lokacii ob"ektov, raspolozhennykh v predelakh atmosfery Zemli
Geod. i fotogramm. — Rostov n/D. (1981). — p. 20—26 : 2 lit.
Improvement of the accuracy of acconting for meteorological parameters in the optical location of objects situated within the Earth's atmosphere (RUS)
AS 12.15.33
Key words: Earth's atmosphere, geodetic measurement

Dobrokhotov, National Correspondent, USSR IGD 84/I 402

528.385.1.024.4

Marushhak, M. P.
O povyshenii tochnosti trigonometricheskogo nivelirovanija iz serediny po vertikal'nym bazisam
Geod. i Kartogr. — Moskva (1982) 10. — p. 25—26 : 4 lit.
On improving the accuracy of trigonometric levelling using vertical bases
(RUS)
AS 12.15.33
Key words: Trigonometric levelling

Dobrokhotov, National Correspondent, USSR IGD 84/I 403

528.27.081.3(44)

Ogier, M. / Sakuma, A.
Mesures absolues de pesanteur en France
Bull. Inform. – Toulouse (1983) 53. – p. 98–113 : 6 fig. 7 tab.
Absolute gravity measurements in France (FRE)
AS 12.15.33/12.15.09
Key words: France, gravity measurement, gravity network

The B.R.G.M. in joint venture with the B.I.P.M. and the French Bureau National de Metrologie has undertaken a national absolute gravity campaign during the February–July period.

Eight absolute gravity measurements located at five new stations (Orleans, Toulouse, Marseille, Dijon and Nancy) have been performed. These stations including Sevres-B.I.P.M. are the fundamental base-stations of the new french gravity network.

The accuracy obtained for the different stations runs from 3 to 8 μGal. This accuracy allows us to perform some scientific observations. The main result is a strong correlation between the gravity and barometric variations.

Another set of experiments is in progress on the pillar of B.R.G.M. scientific office in Orleans in order to define the influences of the water table variations on the absolute gravity measurements.

(Author's summ.)

Techn. Univ. Dresden IGD 84/I 404

528.024.6.061.6

Pazourek, Jiri
Zur Frage der Temperaturkorrektionen von Ergebnissen des hydrostatischen Nivellements
Vermessungstechnik. – Berlin 32 (1984) 2. – p. 57–59 : 1 tab. 10 lit.
The problem of temperature corrections of results of hydrostatic levellings (GER)
AS 12.15.33/12.15.37
Key words: Hydrostatic levelling, temperature correction

In hydrostatic levellings the effect of temperature-variations on two different high water columns on the scales is a main error source for the determination of the difference in height. Discussion of the suitability of temperature corrections for the height of the water level and suggestions for the measuring arrangement without temperature corrections are made.

(Author's summ.)

Techn. Univ. Dresden IGD 84/I 405

528.022.1:528.037

Razumovskaja, T. P.
Issledovanie vlijanija pogreshnostejj centrirovanija na tochnost' izmerenija gorizontal'nykh uglov
Geod. i fotogramm. v gorn. dele. – Sverdlovsk (1982) 9. – p. 22–30 : 3 lit.
Investigation of the effect of centering errors on the accuracy of measuring the horizontal angles (RUS)
AS 12.15.33
Key words: Centering error, horizontal angle

Dobrokhotov, National Correspondent, USSR IGD 84/I 406

528.021.7.088.24

Stanislavichjus, R. B.
K voprosu povyshenija tochnosti svetodal'nomernykh izmerenijj
Geod. i Kartogr. — Moskva (1982) 9. — p. 18–20
On improvement of accuracy of electrooptic distance measurements (RUS)
AS 12.15.33/73.07
Key words: Electrooptical distance measurement, electrooptical range finder

A brief analysis of the methods of determining the mean integral refraction index of the atmosphere based on zenith distance measurements is given. Because of contradictory data on efficiency of such methods the experimental checks on a special calibration geodetic net have been performed. From the results of the studies carried out on 30 sides with length of 9.8 to 17.1 km it is concluded that the said methods are not efficient.

Dobrokhotov, National Correspondent, USSR IGD 84/I 407

528.089.62:535.61-15

Vesely, Miloslav / Vitasek, Josef
Beobachtung von Aenderungen des Richtungsstrahls mittels eines photoelektrischen Nivelliers
Vermessungstechnik. — Berlin 32 (1984) 3. — p. 85–86 : 3 fig. 3 lit.
Observation of variations in the directed beam by a photoelectric level (GER)
AS 12.15.33/73.19
Key words: Atmospheric influence, photoelectric comparator

With a device called "photoelectric comparator" the infrared, almost monochromatic incoherent radiation, generated by a transmitter, is changed into electric power. Its fluctuations permit conclusions with respect to directed beam variations due to the atmosphere. The disadvantage that the deflection of radiation can be observed only at one place of the directions straight line, is partially compensated by the continuous, automatic registration of measuring data.

(Author's summ.)

Techn. Univ. Dresden IGD 84/I 408

AIDOS Systematics **12.15.37 Group I Geodesy**

Mathematical treatment of geodetic measurements

Theory and methods of the mathematical treatment of geodetic measurements
Theory of observation errors and method of least squares
Adjustment and catalogue of the results of measurement
Geodetic calculations
Modern techniques of mathematics and computation for geodetic calculations

Gruppe I Höhere Geodäsie

Mathematische Bearbeitung geodätischer Messungen

Theorie und Methoden der mathematischen Bearbeitung geodätischer Messungen
Theorie der Meßfehler und Methode der kleinsten Quadrate
Ausgleichung und Katalogisierung der Meßergebnisse
Geodätische Berechnungen
Moderne Verfahren der Mathematik und Rechentechnik für geodätische Berechnungen

528.14:519.216.3

Alpár, Gyula
Application of least squares collocation for mineral reserve estimation (ENG)
Acta geod. geophys. montan. Acad. Sci. Hung. – Budapest 16 (1981) 1. – p. 41–49 : 2 fig. 7 lit.
AS 12.15.37/12.15.07/12.23.25
Key words: Least squares collocation, physical geodesy

The planning of investments in mining needs a more and more exact estimation of mineral reserves. For this purpose the so-called "kriging" method based on mathematical statistics is widely applied.

In case of similar estimation problems the least squares collocation method has been successfully applied in physical geodesy.

In the paper the congruence of the two methods is numerically and algebraically proved. The estimation of ore reserves by the least squares collocation method is not only equivalent to kriging, but also offers some remarkable advantages in computing techniques.

(Author's summ.)

Techn. Univ. Dresden IGD 84/I 409

528.1:528.021

Boehm, Josef
Fehler- und Kovarianzanalyse der Streckenmessung in allen Kombinationen mit Bestimmung der Additionskonstante
Vermessungstechnik. – Berlin 32 (1984) 2. – p. 53–54 : 1 tab. 1 lit.
Error and covariance analysis of distance measurement in all combinations with the determination of the addition constants (GER)
AS 12.15.37/12.15.33
Key words: Addition constant, adjustment, distance measurement

Description of the adjustment process and generalization of forms. Error and covariance analysis for the explanation of unfavourable accumulation of errors and of a rare mechanism of the least-squares method. Conclusions: Instead of determining the addition constant by means of adjustment of measurements in all combinations it should be found by a test baseline with exactly known distances.

(Author's summ.)

Techn. Univ. Dresden IGD 84/I 410

519.5
528.59:681.3.02

Deme, Gyula
A geokod elmelete es funkcioi
Geodez. es Kartogr. – Budapest 35 (1983) 3. – p. 164–171 : 5 fig. 15 lit.
A theory of the geocode and some of its functions (HUN)
AS 12.15.37/63.11
Key words: Automatic map drawing, data base, geocode, Urison-formation

The map drawn by a mapping plotter is treated in this paper as a delineation consisting of a single and continuous map line represented by a continuous character series which controls the drawing head of the plotter. Essentially, one should make use of this character series such as of a data base made for general surveying purposes (i.e. are calculation, presentation of different map details etc.), because of mostly economical considerations. Therefore, within this continuous character series depicted on the map by a single continuous line i.e. by a one-dimensional geometric formation, lesser parts should be separated. On the basis of the dimension theory elaborated by a Russian mathematician Urison (1898–

1924), the author finds the most favourable separating formation (named Urison-formation).

Veroe-Hetenyi, National Correspondent, Hungary IGD 84/I 411

519.271/272:528.1

Detrekoei, Akos
A geodeziai es fotogrammetriai meresek sztochasztikus modelljei
Geodez. es Kartogr. – Budapest 35 (1983) 3. – p. 156–159 : 1 fig. 18 lit.
The stochastic model of the geodetic and photogrammetric measurements (HUN)
AS 12.15.37/12.29.25
Key words: Blunder, photogrammetric model, random error, stochastic model, systematic error

This paper discusses the historical background and the development of the stochastic models of the geodetic and photogrammetric measurements. Distinction is made between the models defined by random variables and the ones defined by random functions. For the special kinds of error (blunder, systematic error, random error) the possibility of their "a priori" and "a posteriori" characterization is discussed.

Veroe-Hetenyi, National Correspondent, Hungary IGD 84/I 412

528.1.022

Gan'shin, V. N.
Obshhijj pokhod k obrabotke mnogokratnykh nabljudenijj
Izvest. vyssh. uchebn. Zaved., Geodez. i Aehrofotos"jomka. – Moskva (1982) 4. – p. 3–7 : 7 lit.
A general approach to treatment of repeated measurements (RUS)
AS 12.15.37
Key words: Cyclic measurement

Dobrokhotov, National Correspondent, USSR IGD 84/I 413

528.33:528.1

Gergely, Jozsef
Geodeziai halozatok kiegyenlitese
Geodez. es Kartogr. – Budapest 35 (1983) 1. – p. 18–21 : 1 fig. 7 lit.
Adjustment of geodetic networks (HUN)
AS 12.15.37/12.15.25
Key words: Conjugate gradient method, network adjustment

The adjustment of geodetic networks is a large-scale nonlinear task. In practice the linearized approximations of the nonlinear system of equations are used which requires the solution of the large-size sparse system of the linear equations. In the paper we derive the linearized equations, discuss those solutions by means of conjugate gradient method and formulate the nonlinear task of the network adjustment.

Veroe-Hetenyi, National Correspondent, Hungary IGD 84/I 414

528.38(437):55

Kalvoda, Jan / Zeman, Antonin
Geological interpretation of the results of measurement in 1952–1981 on the test levelling circuit in Prague-Podbaba (ENG)
Sbornik vyzkumn. praci VUGTK, R. 3. – Praha (1982) 14. – p. 89–124 : 16 fig. 1 tab. 23 lit.
AS 12.15.37
Key words: Geology, levelling, movements of the Earth's surface

Geological analysis of the stability of geodetical points by means of repeated levellings on the test levelling circuit in Prague-Podbaba (the 5.5 km long closed poly-

gon with six branch levelling lines, 41 points, 30-year sequence of observations) yields significant information on the vertical component of the present-day movements of the Earth's surface which are of endogenous, exogenous and anthropogenous origins. Geological interpretation of measurements resulted in the fact that the bedrock of the ridges, denudational slopes, flanks and bottoms of river, consisting of schists, greywacke or lydite schists and lydites of proterozoic age are tectonically quiet as their mutual comparisón of movement activity of drilling-, rock outcrops and nailing-rod stabilization has shown. Among the present-day movements of the Earth's surface, movements of exogenous origin, the downwards movements, oscillation and in some places also uplift have been evidenced, which are due to climato-morphogenetic and gravitational processes in slope, fluvial and eolian sediments of quaternary age.
(Author's summ., abbr.)

Techn. Univ. Dresden IGD 84/I 415

528.112:551.32

Koehler, Martin
Zum Einfluss systematischer Fehler bei der Bestimmung lokalen Strains auf Eisoberflaechen
Allg. Vermess.-Nachr. – Karlsruhe 91 (1984) 2. – p. 72–80 : 3 fig. 2 tab. 12 lit.
The effect of systematic errors in the determination of local strain on ice surfaces (GER)
AS 12.15.37/12.15.09
Key words: Strain measurement, systematic error of the strain

In the glaciological literature there are some methods of determining the strain-rate tensor at the surface of an ice shelf or a glacier. They all assume a special shape of deformation figure and the formulas for the strain tensor are very easy to evaluate. In this paper it is shown that in doing so, systematic errors of the principal strains may occur if the deformation figure has not its theoretical shape and if the strain is finite. The effects are demonstrated by a simple example.
(Author's summ.)

Techn. Univ. Dresden IGD 84/I 416

528.1:517.5

Lilienblum, Horst
Genauigkeitsmasse bei Zeitreihenanalysen
Vermessungstechnik. – Berlin 32 (1984) 2. – p. 62–64 : 5 tab. 6 lit.
Accuracy measures in the analysis of time series (GER)
AS 12.15.37
Key words: Analysis of time series, mathematical statistics

Deriving from the necessity to determine accuracy measures of mathematical-statistical parameters of time series, error formulae are compiled and calculation examples are given. The confidence intervals of the linear and quadratic mean, of the variance as well as of the covariance or correlation function of the time series are determined with consideration of the optimum relations of expenditure and results.
(Author's summ.)

Techn. Univ. Dresden IGD 84/I 417

528.118:528.38

Ninkov, T. / Schmitt, G.
Eine Methode der Gewichtsoptimierung in geodaetischen Netzen
Allg. Vermess.-Nachr. – Karlsruhe 90 (1983) 6. – p. 216–222 : 3 fig. 2 tab. 6 lit.
A method of weight optimization in geodetic nets (GER)
AS 12.15.37/12.15.25
Key words: Levelling network, weight optimization

A method for a restricted E-optimal design (optimization of the maximal eigenvalue of the cofactor matrix of the unknowns) is described and demonstrated with two levelling networks.

(Author's summ.)

Techn. Univ. Dresden IGD 84/I 418

528.14:519.21

Petrova, Nadezhda
Metoda na najj-malkite kvadrati pri nabljudavani shuchajjni velichini
Vissha Geod. – Sofija (1983) 9. – p. 46–53
The method of least squares in the case of observations of random quantities (BUL)
AS 12.15.37
Key words: Adjustment, least squares method, mathematical geodesy, random quantities

The classical approach to the least squares method of estimating determined parameters is applied to the non-classical case of adjustment of observations on random quantities. The cases of initial relations, corresponding to the adjustment of indirect observations and to the adjustment by conditions, when the mathematical expectations of the observed values are known or unknown, are examined.

(Author's summ.)

Georgiev, National Correspondent of Bulgaria IGD 84/I 419

528.14:519.2

Schaedlich, Martin
Das arithmetische Mittel als nichtstationaeres Praediktionsfilter
Vermessungstechnik. – Berlin 32 (1984) 2. – p. 50–52 : 2 fig. 13 lit.
The arithmetic mean as non-stationary prediction filter (GER)
AS 12.15.37
Key words: Filtering, mathematical geodesy, prediction

Formulation of a non-stationary filter which is sufficient for an arithmetic mean of time- and position-dependently adjusted observations. The concept provides that the problems of linear prediction filtering may be reduced to the adjustment of "direct" and dependent observations by the "least-squares method". Trend analyses and inversions are not necessary. Formulation of the algorithm and of correlations between the algebraic and physical model component. Numerical example and conclusions.

(Author's summ.)

Techn. Univ. Dresden IGD 84/I 420

528.381:551.241:531.3

Totomanov, Ivan N.
An approach to the synthesis of the two main methods of levelling geokinetic networks data processing (ENG)
Vissha Geod. – Sofija (1983) 9. – p. 54–64
AS 12.15.37/12.15.09
Key words: Adjustment, Earth's crustal movement, precise levelling

It is demonstrated that the known model of adjustment of repeated precise levelling, by unknown estimates x for the heights in a common epoch and the velicities y of recent vertical Earth's crustal movements, could be transformed to an equivalent model of two groups of observation equations: in the first group x and y participate and in the second group only y. The separate processing of only second group of equations is just the method largely utilized in practice, of adjustment of relative velocities by which the estimates $\bar{y} \neq y$ are obtained. An additional adjusting procedure to correct, with the aim to obtain the estimates y, is suggested.

(Author's summ.)

Georgiev, National Correspondent of Bulgaria IGD 84/I 421

528.022.2:528.14

Trenkov, Ivan / Gergov, Cvetan
Ustojjchivost i optimalnost na zadachata za izravnenie na vertikalny ahgli
Vissha Geod. – Sofija (1983) 9. – p. 65–72 : 2 fig. 3 tab.
Stability and optimality in the problem of vertical angle adjustment (BUL)
AS 12.15.37/12.15.33
Key words: Adjustment, vertical angle

The condition numbers of the normal matrix, obtained when adjusting vertical angles measured at a station, and taking into consideration the deflections of the vertical, are examined. The conditions of non-determinacy of the problem are obtained. With a view to obtain the best conditioned normal matrix and the respective un-known estimates of highest accuracy, an experimental study for the optimum observations is carried out.

(Author's summ.)

Georgiev, National Correspondent of Bulgaria IGD 84/I 422

528.118:528.33

Wimmer, Heinrich
Zur Gewichtsoptimierung einer Funktion der Unbekannten
Z. Vermess.-Wesen. – Stuttgart 108 (1983) 11. – p. 484–495 : div. fig. div. lit.
Weight optimization of a function of unknowns (GER)
AS 12.15.37
Key words: Weight optimization

General formulae for the iterative computation of optimal weights are described, where the precision of several functions of coordinate unknowns may be introduced into the target function. The method is specific to the case of only one function, and is tested on the base extension net Munich. The weights of different iteration steps are compared by means of some precision and reliability criteria. It is shown that the weight optimization (second order design) of special purpose networks may by very useful, but one should be cautious in applying optimal weights for proving the accuracy of the network.

(Author's summ.)

Techn. Univ. Dresden IGD 84/I 423

AIDOS Systematics **73** **Group I Geodesy**

Design, manufacture and proof of instruments

Annotation: Problems of theory, calculation, designing, manufacture, examination and proof of instruments and devices (for applications see the respective sections)

Gruppe I Höhere Geodäsie

Konstruktion, Fertigung und Prüfung von Geräten

Anmerkung: Fragen der Theorie, Berechnung, Projektierung, Fertigung, Untersuchung und Prüfung von Geräten (Anwendungen der Geräte in den entsprechenden Abschnitten)

AIDOS Systematics **73.07** **Group I Geodesy**

Construction and proof of geodetic instruments

Distance measuring instruments
Instruments for measuring angles and directions
Instruments for measuring height and inclination
Plotting tables, combined devices
Instruments for measuring oriented directions
Gyroscopes, compasses
Accessories

Gruppe I Höhere Geodäsie

Konstruktion und Prüfung geodätischer Geräte

Geräte zur Streckenmessung
Geräte zur Winkel- und Richtungsmessung
Geräte zur Höhen- und Neigungsmessung
Kartiertische, Gerätekombinationen
Geräte zur Messung orientierter Richtungen
Kreisel, Bussolen
Zusatzgeräte

528.512.089.6:528.14
Csepregi, Sz.
Az elektrooptikai tavmeroek oesszeado-allandojanak es fazishibajanak meghatarozasa a kiegyenlites modositasaval
Erdeszeti es Faipari Egyetem Koezl. – Sopron (1981) 1. – p. 123–136 : 2 fig. 7 tab. 5 lit.
Determination of the phase error and the additive constant of electro-optical stadia rangefinder by modifying the adjustment (HUN)
AS 73.07/12.15.33
Key words: Additive constant, adjustment modification, electro-optical stadia rangefinder

This paper presents a method of determining the additive constant of electro-optical stadia rangefinders.
The determination is made by measuring the points located along one line in each combination. A general relationship can be obtained by calculating with the method of least squares.
Veroe-Hetenyi, National Correspondent, Hungary IGD 84/I 424

528.517
Hipp, J.
Eine neue Entfernungsmesserfamilie mit Puls-Laufzeit-Messverfahren (FEN 2000)
Allg. Vermess.-Nachr. – Karlsruhe 90 (1983) 11–12. – p. 461–464
A new "family" of range finders with pulse-transit time measuring technique (FEN 2000) (GER)
AS 73.07/12.15.33
Key words: Distance measuring technique, range finders, transit time
Techn. Univ. Dresden IGD 84/I 425

528.5:681.33
Kaiser, I. / Mentes, Gy.
Mikroprocesszor-rendszer geodeziai mueszerek fejlesztesehez
Geod. es Kartogr. – Budapest 35 (1983) 5. – p. 333–336 : 2 fig. 4 lit.
Microprocessor system for development of geodetic instruments (HUN)
AS 73.07/63.11
Key words: Geodetic field instrument, hardware design, microprocessor technics, software design

Due to the development of microprocessor techniques it is necessary to increase the capabilities of geodetic instruments with built-in microprocessors. For hardware and software design and test of such instruments a microprocessor system has been developed in the Geodetic and Geophysical Research Institute of the Hungarian Academy of Sciences. The different parts of the system are very flexible to accomplish various kinds of microcomputers needed in geodetic instruments and are suitable to be built into field instruments, too.

Veroe-Hetenyi, National Correspondent, Hungary IGD 84/I 426

528.531:528.482

Lazar, J. / Omaszta, S. / Pfeffer, F.
A Wild TACHYMAT TC 1 alkalmazasa az UVATERV-nel es a meresi adatok feldolgozasa
Geodez. es Kartogr. — Budapest 35 (1983) 1. — p. 38—43 : 3 fig. 3 tab. 5 lit.
Adaption of the Wild TACHYMAT TC 1 at the UVATERV and processing of the measurement data (HUN)
AS 73.07/12.23.21
Key words: Electronic registration tachymeter, map construction, plotting program, Wild Tachymat TC 1

A Wild Tachymat TC 1 electronic registration tachymeter have been in use at the state company Uvaterv since 1980. It can be equally successfully used both in the planning and the execution phase. At present the processing of the data is carried out partly in manual way. The plotting program presented briefly and illustrated by an example is intended to update the data processing.

Veroe-Hetenyi, National Correspondent, Hungary IGD 84/I 427

528.517.089.6

Matsumoto, Shigeo et al.
Instrumental offsets of distance measuring instruments (AGA 8, AGA 700 and K & E Rangemaster III) (JAP)
Bull. Earthq. Res. Inst. — Tokyo 58 (1983) 4. — p. 465—475 : 6 fig. 4 tab. div. lit.
AS 73.07
Key words: Distance measuring instruments

In order to examine instrumental differences of distance measuring instruments, we conducted simultaneous comparative measurements using the AGA 8, AGA 700 and the newly introduced K & E Rangemaster III. Several base-lines and baseline networks with distances from 50 m to 6 km were used for measurements under various conditions. The results suggest that the AGA 700 gives longer distances than the Rangemaster III by about 10 mm depending on the base-line lengths. The longer the base-line, the larger the offset. On the other hand, the AGA 8 gives shorter distances than the Rangemaster III by about 10—20 mm.

Since we are planning to replace the AGA 8 and AGA 700 with the K & E Rangemaster III, these comparative measurements are needed and the obtained instrumental differences should be corrected by connecting the old data of the AGA 700 and AGA 8 with the new data of the K & E Rangemaster III.

Nagasawa, National Correspondent, Japan IGD 84/I 428

528.541.82.089.6

Maurer, Wolfgang / Schnaedelbach, Klaus
First experiences with a vertical comparator for the calibration of invar rods (ENG)
Presented paper to the Workshop on Precise Levelling, March 16—18, 1983, Hannover, F.R.G. — p. 1—8 : 3 fig. 2 tab. 10 lit.
AS 73.07/12.15.33
Key words: Calibration, interferometry, levelling rod

In this report, the performance of a new vertical comparator for an automatic calibration of line-graduated levelling rods by interferometric means is described. Results of comparative calibrations of three pairs are discussed.

(Author's summ.)

Techn. Univ. Dresden IGD 84/I 429

528.517.089.5

Meisenheimer, Dieter
Marktuebersicht ueber EDM-Geraetesysteme
Vermessungsingenieur. — Wiesbaden 34 (1983) 5. — p. 152—162 : 6 tab. 4 lit.
Market report on EDM instrumental systems (GER)
AS 73.07
Key words: Electro-optical distance meter

An outline is given of 40 electro-optical distance measuring instruments (EDM) subdivided into 7 groups in tables.

Techn. Univ. Dresden IGD 84/I 430

528.546:528.024.6

Mozgov, A. K.
Obosnovanie konstrukcijj nekotorykh geodezicheskikh priborov dlja gidrostaticheskogo nivelirovanija
Geofiz. i geodez. metody i sredstva pri poiskakh polez. iskopaemykh v Sibiri. — Novosibirsk (1982). — p. 111—115
Substantiation of the design of some geodetic instruments for hydrostatic levelling (RUS)
AS 73.07/12.15.33
Key words: Hydrostatic levelling

Dobrokhotov, National Correspondent, USSR IGD 84/I 431

528.521.2-187.4:621.38

Muench, K. H.
Der elektronische Praezisions-Theodolit Kern E2
Vermessungsingenieur. — Wiesbaden 34 (1983) 6. — p. 203—207 : 7 fig.
The electronic theodolite of high precision Kern E2 (GER)
AS 73.07
Key words: Theodolite Kern E2

Techn. Univ. Dresden IGD 84/I 432

528.521.089.6

Orban, Aladar
Laboratoriumi modszerek a teodolitok minositesere
Kandidat. ertekezes. — Sopron (1982). — p. 191 : 40 fig. 8 tab.
Laboratory methods for the qualification of theodolites (HUN)
AS 73.07/12.15.33
Key words: Instrumental errors, method for instrument testing, theodolite, tools for instrument testing

Laboratory methods for testing theodolites in different accuracy classes are discussed including technical requirements of a surveying character for these instruments and the tools used in the test. Due to a splitting of errors of different origin both measurement and interpretation are simplified. Some remarks concern the error theory of theodolites. Candidate's thesis.

Veroe-Hetenyi, National Correspondent, Hungary IGD 84/I 433

528.531.089.6

Rohde, Gerd
Untersuchung des elektrooptischen Sekundentachymeters Elta 2 der Fa. Opton Feintechnik GmbH
Vermessungstechnik. — Berlin 32 (1984) 3. — p. 89 : 1 fig.
Investigation of the electro-optical second tachymeter Elta 2 of the firm Opton Feintechnik GmbH (GER)
AS 73.07
Key words: Electro-optical tachymeter Elta 2

Techn. Univ. Dresden IGD 84/I 434

528.5"71"
528.722"71"

Schoeler, Horst
Geodaetische und photogrammetrische Geraete — Tendenzen und Moeglichkeiten aus applikationstechnischer Sicht
Vermessungstechnik. — Berlin 32 (1984) 3. — p. 74—76
Geodetic and photogrammetric instruments — tendencies and possibilities with respect to application (GER)
AS 73.07/73.11
Key words: Geodetic instruments, photogrammetric instruments

The technique of the registration of primary data, of information processing and of the representation of working results must achieve a continuously increasing level to correspond to the demands for mapping of greater and greater scales, for revision of existing map works, for continuous supervision of the environment and for studying geodynamic processes. Deriving from the requirements of national economy as a standard of instrumental technology, some trends in the fields of space technique, microelectronic computer technique, photogrammetry and remote sensing are shown.

(Author's summ.)

Techn. Univ. Dresden IGD 84/I 435

528.531.089.6

Schubert, Roswitha
Untersuchung des Tachymeterautomaten RECOTA des VEB Carl Zeiss JENA
Vermessungstechnik. — Berlin 32 (1984) 3. — p. 88 : 1 fig.
Investigation of the automatic tachymeter RECOTA of the VEB Carl Zeiss JENA (GER)
AS 73.07
Key words: Automatic tachymeter RECOTA

Techn. Univ. Dresden IGD 84/I 436

528.521.2-187.4.088.2

Zahradnik, D.
Die Untersuchung des sekundaeren Kippachsfehlers des Universaltheodolits WILD T4 mit Hilfe eines Laserinterferometers
Z. Vermess.-Wesen. — Stuttgart 109 (1984) 2. — p. 87—92 : 6 fig. 4 lit.
Investigation of the second order error of the horizontal axis of the universal theodolite WILD T4 by means of a laser interferometer (GER)
AS 73.07
Key words: Theodolite axis error, laser-interferometer, universal instrument

The knowledge of the second order error of the horizontal axis of the instrument WILD T4 is essential for precise measurements (e.g. astronomical position fixing).

A new method is presented, which by means of a laser-interferometer allows to measure this second order error without contact and with high resolutions.
(Author's summ.)
Techn. Univ. Dresden IGD 84/I 437

AIDOS Systematics **73.19** **Group I Geodesy**

Geophysical instruments

- for meteorology and oceanography
- for determining the gravity field and the magnetic field
- for seismology

Gruppe I Höhere Geodäsie

Geophysikalische Geräte

- für Meteorologie und Ozeanographie
- für die Bestimmung des Schwere- und Magnetfeldes
- für die Seismik

528.526.6:528.1

Abdelhamid, Kamal Hassan
On the adjustment of gyrotheodolite observations (ENG)
Acta Geodaet., Geophys. et Montan. – Budapest 18 (1983) 1–2. – p. 29–46 : 9 fig. 3 tab. 8 lit.
AS 73.19/12.15.37
Key words: Adjustment of gyrotheodolites, amplitude method, damping, drift of gyrotheodolites, gyrotheodolites, MOM Gi-B 2, reversion point

The paper presents the results of the adjustment calculations for the observations of a gyrotheodolite type MOM Gi-B 2. The purpose of the study was to evaluate some aspects of different orders of best fitting curves to the observations. Three sets of reversion points observed with the three measuring techniques of the amplitude method were used in the measurements. The three sets of observations were adjusted to get the order of the curve which shows the best fitting for both the left and the right reversion points of different sets with minimum corrections. The means of the gyroscope swings were calculated from the adjustment results of both the right and the left reversion points. The damping factor was also calculated from the same results on the basis of the theoretically exact equation of a damped simple harmonic motion which decribes the damping factor.
Veroe-Hetenyi, National Correspondent, Hungary IGD 84/I 438

528.526.6

Abdelhamid, Kamal Hassan
Evaluation of the drift in the gyrotheodolite observations and its influence on gyro-azimuths of the amplitude method (ENG)
Acta Geodaet., Geophys. et Montan. – Budapest 18 (1983) 1–2. – p. 47–58 : 4 fig. 4 tab. 10 lit.
AS 73.19
Key words: Amplitude method, azimuth determination, drift of gyrotheodolites, geodetic instrument, gyrotheodolites, MOM Gi-B 2

The gyrotheodolite used in the investigation is an old one of the type MOM Gi-B 2. The amplitude method with its three measuring techniques was applied. The investigation was made for the data obtained from the adjustment of three observation sets, each consisting of 56 reversion points. For each set, the rest position of the oscillations was calculated and reduced to the instant of release where no drift is yet active, and again calculated as the Schuler mean of the last three adjusted reversion points where the cumulative drift in the observations is greatest. The two values of the rest position were then compared to calculate the drift in the rest position of the swings. The gyro-azimuths were calculated using the two values of the rest position to evaluate the influence of the drift on the azimuths. The results are tabulated and illustrated with diagrams.

Veroe-Hetenyi, National Correspondent, Hungary IGD 84/I 439

528.563

Borodko, A. V. / Lukashuk, S. M. / Neberov, P. A.
Ob uchjote smeshhenija nul'-punktov gravimetrov
Geod. i Kartogr. – Moskva (1983) 1. – p. 39–41
On accounting for zero-drift of the gravimeters (RUS)
AS 73.19/12.15.33
Key words: Gravimeter

Dobrokhotov, National Correspondent, USSR IGD 84/I 440

528.568.089.6:528.27.081.3

Bulanzhe, Ju. D.
Predvaritel'nye rezul'taty sravnenija absoljutnykh gravimetrov
Geod. i Kartogr. – Moskva (1982) 11. – p. 11–15
Preliminary results of comparison of absolute gravity meters (RUS)
AS 73.19/12.15.09
Key words: Gravity meter

In 1981 to 1982 at Sevres (France) a comparison of five absolute gravity meters of essentially different design has been carried out. From convergence of measurements the R.M.S. error of the absolute value of free fall acceleration is determined. It turned out to be about $10 \cdot 10^{-8}$ m/s^2. At the same time the absolute g-value for the pedestal Sevres A 3 has been determined. It turned out to be $(58 \pm 14) \cdot 10^{-8}$ m/s^2 more than the g-value accepted for this station in the system IGSN-71.

Dobrokhotov, National Correspondent, USSR IGD 84/I 441

528.561

Romanjuk, V. A.
Uprugaja deformacija lezvija nozha majatnika-istochnik sistematicheskikh oshibok otnositel'nykh izmerenijj sily tjazhesti
Dokl. AN SSSR. – Moskva (1982) 1. – p. 71–73
Elastic deformation of the pendulum edge as a source of systematic errors of relative gravity measurements (RUS)
AS 73.19/12.15.33
Key words: Gravity measurement, pendulum instrument

Dobrokhotov, National Correspondent, USSR IGD 84/I 442

International Federation of Surveyors

bibliographia geodaetica

AIDOS Systematics **12.23** **GROUP II SURVEYING AND TOPOGRAPHY**

GRUPPE II ANGEWANDTE GEODÄSIE UND TOPOGRAPHIE

AIDOS Systematics **12.23.05** **Group II Surveying and topography**

Geodetic foundations of topographic and engineering-geodetic work

Establishing of local horizontal and vertical control networks, construction and factory networks
Filling-in of first-order networks for engineering-geodetic tasks
Geodetic determination of tie points and control points
Measurements methods (intersection methods, traversing, detailed triangulation) and evaluation methods for establishing local control networks

Gruppe II Angewandte Geodäsie und Topographie

Geodätische Grundlagen für topographische und ingenieurgeodätische Arbeiten

Anlage von lokalen Lage- und Höhenfestpunktnetzen, Bau- und Werksnetzen
Verdichtung von Landesnetzen für ingenieur-geodätische Aufgaben
Geodätische Bestimmung von Anschluß- und Paßpunkten
Messungsmethoden (Einschneideverfahren, Polygonometrie, Kleintriangulation) und Auswerteverfahren zur Anlage lokaler Festpunktnetze

528.02.06:681.3

Benning, Wilhelm
Zur Auswertung nivellitischer, tachymetrischer und trigonometrischer Messungen mit Hilfe mittlerer Datentechnik – Das Programmsystem ATM
Z. Vermess.-Wesen. – Stuttgart 109 (1984) 1. – p. 19–27 : div. lit.
The evaluation of levellings, tachymetric and trigonometrical measurements by mean data technique. – The Programme System ATM (GER)
AS 12.23.05/12.15.37
Key words: Computer program, control network

A computer program package for the densification of ground control networks was designed and written in FORTRAN IV. The computational process involves means for testing the plausibility of data, the automatic reduction of observational data into Gauss-Krueger conformal projection and the automatic adjustment of approximate coordinates. The program includes statistical tests for the detection of gross errors in observations and it offers a-posteriori-variance estimations of the observations. The reliability of observations and coordinates is given, whereas the net adjustment is done by hyper-matrix-methods and separated into horizontal and vertical net adjustment.

(Author's summ.)

Techn. Univ. Dresden IGD 84/II 107

528.48.088:624.2/.8

Dzjaman, A. D.
Vlijanie zadachi optimizacii tochnosti izgotovlenija, montazha i geodezicheskogo obespechenija stroitel'nykh ob"ektov pri fiksirovannom znachenii srednejj kvadraticheskojj oshibki zamykajushhego zvena
Izvest. vyssh. uchebn. Zaved., Geod. i Aehrofotos"jomka. – Moskva (1983) 2. – p. 47–49 : 5 lit.
The effect of the optimization problem for the accuracy of producing, assembling as well as geodetic control of the construction objects for a fixed value of the R.M.S. error of the closing chain (RUS)
AS 12.23.05/12.15.37
Key words: Engineering geodesy

Dobrokhotov, National Correspondent, USSR IGD 84/II 108

528.412

Reminskijj, A. A.
Edinyjj podckod k resheniju geodezicheskikh zasechek
Geod. i Kartogr. – Moskva (1983) 6. – p. 29–33
A single approach to solution of the geodetic intersection-resection problems (RUS)
AS 12.23.05
Key words: Geodetic intersection

Dobrokhotov, National Correspondent, USSR IGD 84/II 109

528.486:624:330.13

Tupolev, O. V.
Ehkonomicheskijj analiz sposobov razbivki stroitel'nojj geodezicheskojj setki
Geod. i Kartogr. – Moskva (1983) 6. – p. 13–14 : 6 lit.
An economic analysis of the methods of laying-out the geodetic construction grid (RUS)
AS 12.23.05/12.23.21
Key words: Economy, geodetic network, setting out

Dobrokhotov, National Correspondent, USSR IGD 84/II 110

528.486:624

Zhukov, B. N.
K raschjotu tochnosti razbivochnykh rabot i stroitel'nykh setok
Izvest. vyssh. uchebn. Zaved., Geod. i Aehrofotos"jomka. – Moskva (1983) 2. – p. 41–46
On accuracy estimation of layout operations and construction grids (RUS)
AS 12.23.05/12.23.21
Key words: Geodetic network, setting out

Dobrokhotov, National Correspondent, USSR IGD 84/II 111

AIDOS Systematics **12.23.09** **Group II Surveying and topography**

Topographic surveying on land

Obtaining information about the terrain by using geodetic methods (tacheometric surveying, plane table surveying and other methods) and means for field survey and for supplementary measurements
Cadastral surveying

Gruppe II Angewandte Geodäsie und Topographie

Topographische Aufnahme des Festlandes

Gewinnung von Informationen über das Gelände mit geodätischen Methoden (tachymetrische Aufnahme, Meßtischaufnahme und andere Verfahren) und Mittel zur Geländeaufnahme und für Ergänzungsmessungen
Katastervermessung

528.425

Bakanova, V. V.
Krupnomasshtabnye topograficheskie s''jomki
Moskva: Nedra, 1983. – 182 p. : div. fig. 52 lit.
Large-scale topographic surveys (RUS)
AS 12.23.09
Key words: Topography
Dobrokhotov, National Correspondent, USSR IGD 84/II 112

528.48(430.2)"71"

Lang, Herbert
Zu einigen Aspekten der Entwicklung der Ingenieurgeodaesie in der Deutschen Demokratischen Republik
Vermessungstechnik. – Berlin 32 (1984) 3. – p. 76–79 : 16 lit.
Some aspects of the development of engineering geodesy in the German Democratic Republic (GER)
AS 12.23.09/12.23.13
Key words: Engineering geodesy

Analysis of some scientific-technological results of recent years and their effect on the increase of the technical-technological level of engineering-geodetic production in the G.D.R. Deriving from the estimation of scientific-technological development of next years on the field of information registration and processing, objects.and their further development are formulated for the determination of horizontal and vertical control points for setting-out and for evidence of deformations.

(Author's summ.)

Techn. Univ. Dresden IGD 84/II 113

528.45(494.28)"47-01"(091)

Sigrist, R.
100 Jahre Stadtvermessung St. Gallen
Vermess. Photogramm. Kulturtechnik. – Zuerich 81 (1983) 11. – p. 389–393 : 5 fig.
100 years – Urban surveying St Gallen (GER)
AS 12.23.09
Key words: Urban surveying
Techn. Univ. Dresden IGD 84/II 114

528.44:347.235

Tegeler, Wilhelm
Moeglichkeiten fuer die Einrichtung eines koordinatenbezogenen Vermessungssystems im Liegenschaftskataster
Z. Vermess.-Wesen. – Stuttgart 108 (1983) 8. – p. 351–359 : 3 tab. 8 lit.
Possibilities to establish a co-ordinate-based surveying system in land registry (GER)
AS 12.23.09
Key words: Cadastre, co-ordinate-based survey system

The four conditions of a co-ordinate-based survey system in land registry are specified. The establishment of a network of very accurate minor control points on the basis of the specifications in Lower Saxony, and the precision and reliability of several methods of object-point- (e.g. boundary points etc.) measurement are discussed. Finally, various possibilities of the gradual establishment of a co-ordinate-based survey system are proposed.
(Author's summ.)
Techn. Univ. Dresden IGD 84/II 115

528.4"71"

Toepfer, Friedrich
Zur Entwicklung der topographischen Landesaufnahme
Vermessungstechnik. – Berlin 32 (1984) 2. – p. 43–46 : 2 fig. 11 lit.
The development of topographic national surveying (GER)
AS 12.23.09
Key words: Topographic surveying

Deriving from present modes of operation of topographic national surveying the effects of automated photogrammetric and geodetic techniques are analyzed and the fundamental concepts of future procedures are derived. Special attention should be paid to the data flow from photogrammetry to cartography and the work with the data bank. Topographic researches for registration of semantic information and completion of detail should be included now as before.
(Author's summ.)
Techn. Univ. Dresden IGD 84/II 116

AIDOS Systematics **12.23.13** **Group II Surveying and topography**

Topographic and engineering-geodetic maps

Content of large-scale topographic and engineering-technical maps and plans and their assignment to definite purposes
Land maps
Symbols and methods of representation
Automation of the production and revision of large-scale topographic, engineering-technical maps and plans as well as of land maps (on the basis of geodetic methods of information acquisition)
Digital terrain models

Gruppe II Angewandte Geodäsie und Topographie

Topographische und ingenieurgeodätische Karten

Inhalt und Zweckbestimmung großmaßstäbiger topographischer und ingenieurtechnischer Karten und Pläne
Liegenschaftskarten
Signaturen und Darstellungsverfahren
Automatisierung der Herstellung und Laufendhaltung großmaßstäbiger topographischer, ingenieurtechnischer Karten und Pläne sowie von Liegenschaftskarten (auf der Grundlage geodätischer Verfahren der Informationsgewinnung)
Digitale Geländemodelle

681.3.05:528
681.327.2:528

Kluge, Waldemar
Ueber die Nutzung digitaler geodaetischer Daten
Vermessungstechnik. – Berlin 32 (1984) 2. – p. 38–40 : 22 lit.
The utilization of digital geodetic data (GER)
AS 12.23.13/12.29.25
Key words: Digital geodetic data

Characterization of drafted, developed, and partially realized EDP-projects referring to information (data) of terrain, topographic objects, cadastral and special thematic items. Reference systems for the co-ordination of the above data are discussed. Tendencies for rationalization and efficient realization of development and utilization of different digital data storage units are described.
(Author's summ.)
Techn. Univ. Dresden IGD 84/II 117

528.932.3:681.3.04
Leberl, F. / Glaenzer, S. / Beer, M.
Herstellung sehr dichter Hoehenraster aus digitalisierten Schichtenlinien
Z. Vermess.-Wesen. – Stuttgart 109 (1984) 1. – p. 27–34 : 3 fig. 4 tab. div. lit.
The production of very dense rasters from digitized contour lines (GER)
AS 12.23.13
Key words: Digital contour line model
This paper describes a procedure to convert analog contour line sheets via raster scanning and vectorisation to digital form, to add height attributes and generate a height raster. Typically, contour sheets of maps 1 : 50,000 are used for national DHMs at contour intervals of 10 or 20 m. Raster mesh size in geographical φ, λ is defined by $\Delta\varphi = \Delta\lambda = 1''$. This leads to $8 \cdot 10^5$ raster points per sheets with accuracies of about $\pm$ 2 m.
(Author's summ.)
Techn. Univ. Dresden IGD 84/II 118

528.425(086.5):681.3
Markus, Bela / Molnar, T.
A domborzat jellegzetessegeinek feltarasa szamitogeppel
Geod. es Kartogr. – Budapest 35 (1983) 4. – p. 279–283 : 5 fig. 5 lit.
Computer-aided investigation of terrain characteristics (HUN)
AS 12.23.13
Key words: Cluster analysis, digital terrain model, sift-and shift method
Veroe-Hetenyi, National Correspondent, Hungary IGD 84/II 119

528.48:681.3
Markus, Bela
Geodezicheskaja baza avtomatizirovannogo proektirovanija inzhenernykh sooruzhenijj
Period. Polytechn. : Ser. Civil Engin. – Budapest 27 (1983) 3–4. – p. 221–226 : 1 fig. 14 lit.
Geodetic background of CAD in civil engineering (RUS)
AS 12.23.13/12.29.25
Key words: Automatic design, computer aided design, digital terrain model
Geodesy is expected to provide terrain data accessible to a computer for CAD, as well as service programs for evaluating the data in cases where the terrain is decisive for the design process. The resulting digital relief model facilitates spatial design, i.e. simultaneous horizontal and elevational design. The likelihood of the rapid extension of CAD requires modification of study programs.
Veroe-Hetenyi, National Correspondent, Hungary IGD 84/II 120

528.45(084.3):528.936
Osipuk, E. S.
Povyshat' tochnost' obnovljaemykh planov gorodov
Geod. i Kartogr. – Moskva (1983) 4. – p. 26–28
Increasing the accuracy of the city plans to be up-dated (Experience of a geodetic production agency) (RUS)
AS 12.23.13/12.23.09
Key words: Revision of city plan
Dobrokhotov, National Correspondent, USSR IGD 84/II 121

528.425(086.5):517.518.45

Vovk, I. G. / Kostyna, Ju. G. / Suzdalev, A. S.
Matematicheskoe modelirovanie rel'efa uchastka mestnosti rjadami Fur'e
Probl. avtomatiz. topogr.-geod. i kartogr. rabot. Materialy Vses. nauch.-tekhn. konf., Novosibirsk, 26—28 maja, 1981. — Moskva (1982). — p. 60—63
Mathematical modeling of the relief of a limited area by Fourier series (RUS)
AS 12.23.13/12.15.37
Key words: Digital terrain model, Fourier series

Under supposition that the relief is a continuous function in the area of modeling it is suggested to approximate it by the generalized Fourier series. As the series is convergent only within the area it is considered as expedient to widen the area so that the first derivatives of the relief function should remain continuous and to perform approximation in this widened area. The iterative method with estimation of deviations at the basic points after each iteration is described. If the errors become non-admissible ones the process goes on but each time height differences at basic points are approximated.

Dobrokhotov, National Correspondent, USSR IGD 84/II 122

AIDOS Systematics **12.23.17 Group II Surveying and topography**

Topographic surveying of the sea bottom

Topographic and thematic surveying of the shelf, the ocean bottom and the sea bottom
Content of sea bottom maps
Coastal and sea surveying
Methods and means for determining the co-ordinates of observation stations and measured objects

Gruppe II Angewandte Geodäsie und Topographie

Topographische Aufnahme des Meeresbodens

Topographische und thematische Aufnahme des Schelfs, des Ozean- und Meeresbodens
Inhalt von Karten des Meeresbodens
Küsten- und Seevermessung
Methoden und Mittel zur Bestimmung der Koordinaten von Meß- und Aufnahmepunkten

528.47

Emel'janov, O. G.
Obobshhenie rezul'tatov s"jomki rel'efa morskogo dna
Geod. i Kartogr. — Moskva (1983) 5. — p. 38—41 : 4 lit.
Generalization of the results of sea-bottom survey (RUS)
AS 12.23.17/12.15.21
Key words: Marine geodesy, shelf

Dobrokhotov, National Correspondent, USSR IGD 84/II 123

528.47

Kamornyjj, V. M. / Nikolaenko, A. I.
Opyt kompleksnojj topograficheskojj s"jomki shel'fa
Geod. i Kartogr. — Moskva (1983) 5. — p. 25—28
Experience with the integrated topographic survey of the shelf (RUS)
AS 12.23.17/12.15.21
Key words: Marine geodesy, shelf topography

Dobrokhotov, National Correspondent, USSR IGD 84/II 124

528.47:622.242
Kelly, J.
Problems of position-fixing in the magnus field development (ENG)
RICS J. (Land and Minerals Survey.). – London 1 (1983) 2. – p. 69–81 : 8 fig.
AS 12.23.17
Key words: Hydrographic surveying, position-fixing at sea
Cotter, National Correspondent, of Great Britain IGD 84/II 125

828.472:681.883:531.719.35
Kolouch, Dieter
Interferometrisches seitwaerts schauendes Sonar. Ein System zur flaechenhaften topographischen Seevermessung
Z. Vermess.-Wesen. – Stuttgart 108 (1983) 11. – p. 461–470 : 10 fig. 5 lit.
Interferometric side-scan-sonar. A topographic surveying system of coastal areas (GER)
AS 12.23.17
Key words: Coastal engineering, echo-sounder, interferometry
The knowledge of bottom topography of our oceans is one of the assumptions for coastal engineering and marine geology. Especially in shallow coastal areas, however, topography is frequently changed by storms and tidal movements. Because of operational restrictions conventional systems for depth measurements like echo-sounders are not able to take actual data in short-time repetition rates. The development of new systems like the interferometric side-scan-sonar and corresponding digital evaluation methods can solve these problems.
(Author's summ.)
Techn. Univ. Dresden IGD 84/II 126

AIDOS Systematics **12.23.21 Group II Surveying and topography**

Application of geodesy in construction engineering and in traffic engineering

Engineering surveying operations in the designing, construction and using of industrial, civil-engineering, residential and public buildings
Reconnaissance, setting-out and route- and location surveys
Deformation measurements, subsidence and settlement measurements
Methods and means of engineering surveying

Gruppe II Angewandte Geodäsie und Topographie

Anwendung der Geodäsie im Bau- und Verkehrswesen

Ingenieurgeodätische Arbeiten bei der Projektierung, Errichtung und Nutzung von Industrie-, Ingenieur-, Wohn- und Gesellschaftsbauten
Erkundungen, Absteckungs- und Trassierungsarbeiten
Deformationsmessungen, Setzungs- und Senkungsmessungen
Methoden und Mittel der Ingenieurgeodäsie

528.482:627.8(728.1)
Belat, R.
Deformationsmessungen im Rio Chixoy-Projekt
Vermess. Photogramm. Kulturtechnik. – Zuerich 81 (1983) 12. – p. 437–441 : 7 fig.
Deformation measurements in the Rio Chixoy Project (GER)
AS 12.23.21/12.23.29
Key words: Deformation measurement, engineering geodesy, rockfill dam
The present article describes the instrumentation and deformation measurements on the Pueblo Viejo Dam as well as the surveillance of an active geological fault.

It will be explained that in developing countries, which do not dispose of adjustment programmes on large computers or when the customer is not willing to have data processed abroad, good results may be obtained by simple methods.

(Author's summ., abbr.)

Techn. Univ. Dresden IGD 84/II 127

528.482

Detrekoei, Akos
Die Rolle der Zeit in der Planung ingenieurgeodaetischer Deformationsmessungen
Period. Polytechn. : Ser. Civil Engin. – Budapest 27 (1983) 3–4. – p. 125–132 : 16 lit.
Impact of time in planning engineering geodesy deformation tests (GER)
AS 12.23.21/12.23.05/12.15.37
Key words: Engineering geodesy, test measurement organization

After presenting aims and peculiarities of planning engineering geodesy tests, the impact of time on the determination of measurement durations and on the appointment of measurement times will be outlined.

Determination of measurement times by optimum calculus permits to extend the Grafarend system of planning geodesy networks. Optimum planning of measurement times may be that by fourth-order optimum planning.

As concerns the optimum selection of measurement technology and times, a method relying on Legendre polynomials has been suggested, illustrated in an example.

Veroe-Hetenyi, National Correspondent, Hungary IGD 84/II 128

528.48:531.3

Farkas, Ervin / Graczka, Gy.
Testing dynamic motion of a body with a given surface (ENG)
Period. Polytechn. : Ser. Civil Engin. – Budapest 27 (1983) 3–4. – p. 183–187 : 5 fig.
AS 12.23.21
Key words: Dynamic vibration, parabolic aerial laser beam, signal form analyser

Three new measurement methods have been developed for determining and evaluating dynamic motions of bodies (e.g. a parabolic aerial).

First, a laser beam reflected by a mirror on the aerial containing in its direction the dynamic vibrations is received by an active detector. The information on the dynamic vibration is evaluated by a microprocessor signal form analyser. Second, the principle of positive feedback is used by means of vibration transformer pick-up, an amplifier and a loudspeaker. The assembled system is excited at the resonance frequency of the parabole. Third, the information is visualized by means of a quick-run camera recording the laser beam containing information on the dynamic vibration.

Veroe-Hetenyi, National Correspondent, Hungary IGD 84/II 129

528.48.024.4:624.9

Gruzinov, V. V. / Shashkov, V. A.
Peredacha otmetok na vysokie sooruzhenija
Geod. i Kartogr. – Moskva (1983) 7. – p. 27–29
Height determination on elevated constructions (RUS)
AS 12.23.21/12.15.33
Key words: Height transfer

Dobrokhotov, National Correspondent, USSR IGD 84/II 130

528.486.088

Herda, Miroslav
Epitesi tueresek szamitasa es a kituezes pontossaga
Geod. es Kartogr. – Budapest 35 (1983) 4. – p. 263–266 : 1 tab. 6 lit.
Computation of building tolerances and accuracy of layout (HUN)
AS 12.23.21
Key words: Engineering geodesy, layout, limit dimensions, probability

Veroe-Hetenyi, National Correspondent, Hungary IGD 84/II 131

528.482

Horvath, Kalman / L'Aune, O.
70 m magas kemeny elmozdulasvizsgalata linearis eloemetszessel
Geod. es Kartogr. – Budapest 35 (1983) 4. – p. 258–263 : 5 fig. 18 lit.
The measurement of displacement of a 70 m high chimney by linear intersection (HUN)
AS 12.23.21
Key words: Engineering geodesy, intersection, leaning of industrial structures

Results of precise geodetic measurements demonstrate the tested industrial structure – chimney stack – to have leaned because of technically insufficiently detectable causes (foundation or construction deficiencies; one-sided atmospheric effects, etc.).

The regular displacement was by orders exceeded by that due to atmospheric effects.

The expected displacement value – outside the test period – being inaccessible to extrapolation, it may be the intensity of leaning during unit time would still increase – no reliable opinion on the displacement value is possible other than by a systematic measurement series.

Veroe-Hetenyi, National Correspondent, Hungary IGD 84/II 132

528.512.089.6:531.714.1
528.48.02:521.039.5

Horvath, Kalman / L'Aune, O.
Praezisionslaengenbestimmung von Komparatorbasen in einem Kernkraftwerk im Bau
Period. Polytechn. : Ser. Civil Engin. – Budapest 27 (1983) 3–4. –
p. 133–147 : 7 fig. 1 tab. 18 lit.
Precise determination of comparator basis lengths in a nuclear power station under construction (GER)
AS 12.23.21
Key words: Comparator basis length, nuclear power station, statistical calibration

Lengths of five comparator bases of common origin but different lengths on the site of a nuclear power station now under construction have been precisely determined in order to calibrate measuring tapes used in the design and construction of the nuclear power station according to the same unit – "international standard meter" – at an accuracy of $\pm$ 0.2 mm.

High-precision length determination of the pair of standard meters applied in measurements relied on the method of so-called "statistical calibration", complemented by applications of a laser interferometer and of a universal measuring device assisted technically by the National Office of Metrology.

Veroe-Hetenyi, National Correspondent, Hungary IGD 84/II 133

528.48

Kiss, Albert
Planirovanie geodezicheskikh rabot pri montazhe stal'nykh konstrukcijj
Period. Polytechn. : Ser. Civil Engin. — Budapest 27 (1983) 3—4. —
p. 197—208 : 1 fig. 11 lit.
Planning geodetic works for the assembly of steel structures (ENG)
AS 12.23.21
Key words: Assembly of steel structures, engineering geodesy

Up-to-date geodetic facilities and the applied geodetic technology have a great significance and economical impact for the design and assembly of steel structures, in determining the role of geodetic work in design, and the design tasks of the geodesist. Recently, the Institute of Geodesy of the Technical University, Budapest undertook geodetic works for some important constructions with steel structure, observations made in, and publications are relied on in analyzing geodetic tasks and duties of the geodesist in assembling steel structures, illustrated on the example of a geodetic assembly control undertaken by this Institute.

Veroe-Hetenyi, National Correspondent, Hungary IGD 84/II 134

528.02:528.482

Krauter, Andras
Opredelenie formy i deformacii dannojj poverkhnosti geodezicheskim metodom
Period. Polytechn. : Ser. Civil Engin. — Budapest 27 (1983) 3—4. —
p. 209—220 : 7 fig. 3 tab.
Geodetic method for determining form and deformation of a given surface (RUS)
AS 12.23.21/12.15.33
Key words: Deformation of surfaces, form of surfaces, manufacturing accuracy, microwave antenna

Checking the manufacturing accuracy, and testing the deformation under thermal and static loads of the surface of a microwave telecommunication antenna will be described, with the planning of measurements, the expected accuracy, and evaluation of the confidence of results. Data referring to the shape, position and orientation of the quadratic paraboloid of rotation best fitting points of the antenna surface may be determined by the least squares method using a computer. An approximate method suiting determination of form and characteristics of the fitting surface using a small desk-top calculator is suggested.

Veroe-Hetenyi, National Correspondent, Hungary IGD 84/II 135

621.375:528.48

Maurer, Bernhard
Einsatz eines Rundum-Lasers in extremen Neigungen
Vermessungsingenieur. — Wiesbaden 34 (1983) 6. — p. 210—212 : 2 fig.
The application of a roundabout laser in extreme inclinations (GER)
AS 12.23.21/73.07
Key words: Laser instrument

Techn. Univ. Dresden IGD 84/II 136

528.48:711.4

Odor, Karoly
Standortplanung und geodaetische Projekte von Industrieanlagen
Period. Polytechn. : Ser. Civil Engin. — Budapest 27 (1983) 3—4. —
p. 157—169 : 4 fig.
Layout and geodetic design of industrial plants (GER)
AS 12.23.21/83
Key words: Design of industrial plant, geodetic design, layout of industrial plant

The layout design of industrial plants involves a number of problems to be solved by geodesists such as:

- to reconcile construction plans and siting plans
- determination of co-ordinates of the position of planned constructions
- drawing the master plan and the set-out plans.

A unified concept is given of the design work concerned with the layout of constructions, stressing geodetic design problems.

Veroe-Hetenyi, National Correspondent, Hungary IGD 84/II 137

681.327.2:711

Rehatschek, Juergen
Digitale topographische Informationen fuer die automatengestuetzte Bauplanung und Bauvorbereitung innerstaedtischer Raeume
Vermessungstechnik. – Berlin 32 (1984) 3. – p. 79–82 : 3 fig. 4 lit.
Digital topographic data for computer-assisted constructional planning and preparation of building projects of inner urban areas (GER)
AS 12.23.21/12.23.13
Key words: Digital topographic data, land information system

Inner urban construction needs a highly developed organization of data aquisition and processing. An effective mechanical basic system for constructional planning and preparation processes is the data storage "housing policy". The area code file of the data storage should ensure the incorporation into a land information system, which is to be created, by means of local co-ordinates and a uniform grid system.

(Author's summ.)

Techn. Univ. Dresden IGD 84/II 138

528.482:624.19
528.022.1/.2

Rossipal, Gunter
Konvergenzmessungen durch Winkelmessungen
Vermessungsingenieur. – Wiesbaden 34 (1983) 5. – p. 166–170 : 7 fig. 2 lit.
Convergence measurements by angle measurements (GER)
AS 12.23.21
Key words: Angle measurement, tunnel surveying

Techn. Univ. Dresden IGD 84/II 139

AIDOS Systematics **12.23.25** **Group II Surveying and topography**

Application of geodesy in mining

Methods and means for carrying out underground-surveying-geodetic work
Mining surveying
Mining cartography

Gruppe II Angewandte Geodäsie und Topographie

Anwendung der Geodäsie im Bergbau

Methoden und Mittel zur Ausführung markscheiderisch-geodätischer Arbeiten
Markscheidewesen
Bergmännisches Rißwerk

528.48:528.1

Funcke, G.
Datumsinvariante Deformationsanalyse mit zwangsfreien Netzen
Allg. Vermess.-Nachr. — Karlsruhe 91 (1984) 2. — p. 80—83 : 2 fig.
Datum-invariant deformation analysis with contraint-free nets (GER)
AS 12.23.25/12.23.21/12.15.37
Key words: Adjustment of geodetic nets, deformation analysis

Techn. Univ. Dresden IGD 84/II 140

528.481.023.3:528.531:681.3

Kahmen, Heribert / Suhre, Horst
Ein lernfaehiges tachymetrisches Vermessungssystem zur Ueberwachung kinematischer Vorgaenge ohne Beobachter
Z. Vermess.-Wesen. — Stuttgart 108 (1983) 8. — p. 345—351 : 3 fig. 1 lit.
Tachymetric surveying system which is capable of "learning" for the supervision of kinematic processes without observer (GER)
AS 12.23.25
Key words: Computer-controlled tacheometer, deformation measurement

The development of an automatic computer-controlled tacheometer system is described. The system provides automatic monitoring of an array of slowly moving control points for deformation measurements, manages the recording of the data and computes the displacements. Only the initial pointing on a reference point requires an operator.

(Author's summ.)

Techn. Univ. Dresden IGD 84/II 141

528.481:519.2

Moeser, Michael
Die Kovarianzfunktion einer Hangbewegung
Vermessungstechnik. — Berlin 32 (1984) 2. — p. 55—57 : 2 fig. 5 lit.
The covariance function of a slope movement (GER)
AS 12.23.25/12.15.37
Key words: Variance-covariance function

The variance-covariance matrix is an essential part of the prediction and collocation method. The elements of this matrix are determined by an autocovariance function (ACF). The empirical covariances are determined from remaining residuals after separation of trend. To estimate the elements of signal covariance matrices, the empirical covariances are approximated by a suitable function. For the slip area of an open mine the confidence intervals of the empirical ACF were compared with two model ACF's and their parameters were determined.

(Author's summ.)

Techn. Univ. Dresden IGD 84/II 142

528.481:622.271.3

Nehring, H.
Markscheiderische Deformationsmessungen an Tagebauboeschungen
Z. Vermess.-Wesen. — Stuttgart 108 (1983) 8. — p. 339—345 : 4 fig. 2 lit.
Deformation measurements on open cast mine slopes (GER)
AS 12.23.25
Key words: Deformation measurement, open cast

With regard to the special conditions prevailing in large open cast mines the possibilities — available in recent times — of adjusting geodetic networks by hypothesis tests, calculating photogrammetric blocks as well as electrooptical distance measurement were further developed in terms of methodology and instrumentation, so that the ground movements of the open cast mine slopes can be controlled by the mine surveyors reliably and in time.

(Author's summ.)

Techn. Univ. Dresden IGD 84/II 143

AIDOS Systematics **12.23.29** **Group II Surveying and topography**

Application of geodesy in agriculture, forestry and water resources management and for environmental protection

Geodetic work for land management and the preparation of large-scale ground maps, forestry maps and water-resources-management maps
Ground cadastre
Forest taxation

Gruppe II Angewandte Geodäsie und Topographie

Anwendung der Geodäsie in der Land-, Forst- und Wasserwirtschaft sowie beim Umweltschutz

Geodätische Arbeiten für die Landeinrichtung und Herstellung von großmaßstäbigen Boden-, Forst- und wasserwirtschaftlichen Karten
Bodenkataster
Forsttaxation

528.46:626.8

Abuev, O.
Sovershenstvovanie inzhenerno-geodezicheskikh rabot dlja meliorativnogo stroitel'stva v Karakalpakskojj ASSR
Vestn. Karakalp. fil. AN UzSSR. — (1983) 1. — p. 32—34 : 7 lit.
Improvement of engineering geodetic operations for amelioration construction in the Karakalpak ASSR (RUS)
AS 12.23.29
Key words: Amelioration, engineering geodesy

Dobrokhotov, National Correspondent, USSR IGD 84/II 144

528.46:626.81/.85

Emec, E. I.
O vybore masshtaba plana i sechenija rel'efa pri proektirovanii poverkhnostnogo oroshenija
Inzh. geod. — Kiev (1983) 26. — p. 8—10 : 3 lit.
On determination of the scale and the relief contour interval of a plan for designing the surface irrigation (RUS)
AS 12.23.29
Key words: Engineering geodesy, irrigation

Dobrokhotov, National Correspondent, USSR IGD 84/II 145

528.44:711.16

Gastaldi, Jacques
Le remembrement-amenagement doit etre l'oeuvre conjointe du geometre rual et de l'urbaniste
Geometre. — Paris 126 (1983) 11. — p. 59—60
Regional planning and land improvement must be a common project of surveyors and municipal architects (GER)
AS 12.23.29
Key words: Land improvement, regional planning

Report for the FIG-Commission 7.

Techn. Univ. Dresden IGD 84/II 146

528.46:631/633

Maslov, A. V.
Razvitie i sovershenstvovanie topografo-geodeszicheskikh izyskanijj dlja sel'skogo khozjajjstva
Izvest. vyssh. uchebn. Zaved., Geod. i Aehrofotos"jomka. – Moskva (1983) 3. – p. 11–18
Development and improvement of topographic-geodetical reconnaissance operations for agriculture (RUS)
AS 12.23.29
Key words: Agriculture, engineering geodesy

Dobrokhotov, National Correspondent, USSR IGD 84/II 147

528.482:69.058

Penndorf, Hans-Guenter / Turbing, Manfred
Zur Anwendung des Schwingsaiten-Extensometers SMG 11
Vermessungstechnik. – Berlin 32 (1984) 2. – p. 59–61 : 5 fig. 1 tab. 10 lit.
The application of the vibrating string extensometer SMG 11 (GER)
AS 12.23.29
Key words: Vibration string extensometer

Description of the measuring device its operation mode and the determination of the characteristic line of the combination sensor/surveying wire. Studies of material and accuracy showed that variations in length are measurable with an accuracy of $m_l = 2.7\ \mu m/mm$. Applications are illustrated with an example from the field of rock mechanics and of monitoring a dam.
(Author's summ.)

Techn. Univ. Dresden IGD 84/II 148

528.46:634.0:681.3

Polosin, Ju. K. / Musaev, I. G.
Avtomatizacija obrabotki dannykh inzhenerno-geodezicheskikh izyskanijj lesozagotovitel'nykh predprijatijj
Mekhaniz. lesozagot. i transp. lesa. – Leningrad (1983). – p. 53–56
Automation of the processing of engineering geodetic survey data in timber industry (RUS)
AS 12.23.29
Key words: Engineering geodesy, forestry

Dobrokhotov, National Correspondent, USSR IGD 84/II 149

528.46:626.81/.85

Tumar, A. V.
K voprosu vybora vysoty sechenija rel'efa pri proektirovanii osushenija zakrytym drenazhjom
Izvest. vyssh. uchebn. Zaved., Geod. i Aehrofotos"jomka. – Moskva (1983) 2. – p. 50–54 : 8 lit.
On determination of the relief contour interval in designing the irrigation by closed drainage (RUS)
AS 12.23.29
Key words: Engineering geodesy, irrigation

Dobrokhotov, National Correspondent, USSR IGD 84/II 150

International Union of Geodesy and Geophysics
International Association of Geodesy

bibliographia geodaetica

AIDOS Systematics **12.29** **GROUP III PHOTOGRAMMETRY AND REMOTE SENSING**

GRUPPE III PHOTOGRAMMETRIE UND FERNERKUNDUNG

AIDOS Systematics **12.29.05** **Group III Photogrammetry and remote sensing**

Fundamentals of photogrammetry and remote sensing

Mathematical basis of photogrammetry
Affine and projective descriptive geometry for photogrammetric processes
Optical and photogrammetric imaging
Stereoscopic vision and measurement
Types and properties of photograms

Gruppe III Photogrammetrie und Fernerkundung

Grundlagen der Photogrammetrie und Fernerkundung

Mathematische Grundlagen der Photogrammetrie
Affine und projektive darstellende Geometrie für photogrammetrische Prozesse
Optische und photogrammetrische Abbildungen
Stereoskopisches Sehen und Messen
Arten und Eigenschaften der Meßbilder

528.7-187.4:528.711.11
Burch, J. M. / Forno, C.
Preliminary assessment of a new high precision camera (ENG)
In: Precision and speed in close range photogrammetry : ISPRS Sympos., York, England, 5th to 10th September, 1982. — London, 1982. — (Internat. Archives Photogramm. 24 (1982) V/1). — p. 90—99 : 7 fig. 2 tab.
AS 12.29.05
Key words: Precision camera

A monocentric axicon camera has been designed for use in high precision photogrammetry with steel ball or point source targets. An example has been constructed and assembled successfully in a prototype camera body. The axicon patterns which it records remain sharp and symmetrical for all object distances from 300 mm to infinity, over an angular field 40° in diameter. Theoretically the distortion of this camera should be zero, and first measurements indicate that it does not exceed 100 nm across 50 mm plate-width.
(Author's summ.)
Techn. Univ. Dresden IGD 84/III 109

528.711.1(21-194.2):771.531.37
Fraser, C. S.
Film unflatness effects in analytical non-metric photogrammetry (ENG)
In: Precision and speed in close range photogrammetry : ISPRS Sympos., York, England, 5th to 10th September, 1982. — London, 1982. — (Internat. Archives Photogramm. 24 (1982) V/1). — p. 156—166 : 3 fig.
AS 12.29.05/12.29.25/12.29.33
Key words: Analytical photogrammetry, close range photogrammetry, film deformation

A significant problem common to analytical data reduction schemes for non-metric imagery is that the adopted mathematical model is invariably incomplete, especially in relation to the modelling of film deformation. This paper discusses the application of a self-calibration adjustment approach for non-metric photography, which employs a mixed block- and photo-invariant additional parameter concept. The method yields high accuracy, but is still subject to imperfect functional model fidelity due to photographic surface unflatness. This problem can lead to the unfortunate case of high internal consistency and design-level precision, but less than optimum accuracy. A close-range photogrammetric experiment conducted partially in order to assess film unflatness effects on attainable accuracy and precision is outlined, and the results are discussed.
(Author's summ.)
Techn. Univ. Dresden IGD 84/III 110

528.711:629.73(091)

Georgopoulos, A.
Balloon and kite photography: An historical review (ENG)
In: Precision and speed in close range photogrammetry : ISPRS Sympos., York, England, 5th to 10th September, 1982. — London, 1982. — (Internat. Archives Photogramm. 24 (1982) V/1). — p. 196—206 : 3 tab. 62 lit.
AS 12.29.05
Key words: Ballon photography, kite photography

A thorough historical review of low altitude photography experiments, using balloons and kites as camera platforms, is presented. The most important efforts are described and assessed for their results and significance in terms of their photogrammetric applicability.
(Author's summ.)
Techn. Univ. Dresden IGD 84/III 111

528.7(21-194.2):621.375

Pryputniewicz, Ryszard J.
High precision hologrammetry (ENG)
In: Precision and speed in close range photogrammetry : ISPRS Sympos., York, England, 5th to 10th September, 1982. — London, 1982. — (Internat. Archives Photogramm. 24 (1982) V/2). — p. 377—386 : 3 fig. 12 lit.
AS 12.29.05
Key words: Hologram interferometry, hologrammetry

The method of heterodyne hologram interferometry is presented. It is shown that, using this method, displacements can be measured non-invasively with an accuracy of $3 \cdot 10^{-10}$ m, strains — with an accuracy of 0.2 μm/m, and curvature — with an accuracy of $9 \cdot 10^{-5}$ m^{-1}. Representative applications of the method are also discussed, in general. The heterodyne hologram interferometry finds particular applications in displacement and deformation analysis. It is especially suited for determination of objects shaped and sized by contouring. It can also be used in heat transfer and fluid mechanics studies. Today, the heterodyne hologram interferometry is the most precise and accurate method for non-invasive measurements of structures, devices, and processes.
(Author's summ.)
Techn. Univ. Dresden IGD 84/III 112

528.74:621.375

Pryputniewicz, Ryszard J.
Laser speckle method for biostereometrics (ENG)
In: Precision and speed in close range photogrammetry : ISPRS Sympos., York, England, 5th to 10th September, 1982. — London, 1982. — (Internat. Archives Photogramm. 24 (1982) V/2). — p. 387—396 : 7 fig. 12 lit.
AS 12.29.05/12.29.25
Key words: Laser speckle method

Fundamentals of double-exposure laser speckle method are outlined. Application of the method in non-invasive quantitative biostereometric studies is discussed. It is shown that two or more simultaneously recorded specklegrams can be used to determine three-dimensional motions. This analysis, based on the concept of observed displacement vector combined with the concept of projection matrix, allows separation of total object motions into rigid-body translations and rigid-body rotations, which are particularly useful in quantitative studies of force-displacement relationship in biostereometrics.
(Author's summ.)
Techn. Univ. Dresden IGD 84/III 113

528.7

Sears, G. M.
First steps in photogrammetry (ENG)
Dep. Land Survey. (North East London Polytechn.). – Dagenham 5 (1982). – 158 p. : 94 fig. 8 lit.
AS 12.29.05
Key words: Photogrammetry

The study of photogrammetry and definition of the objects.
Cotter, National Correspondent, of Great Britain IGD 84/III 114

528.711.7(100)

. . .
International Journal of Remote Sensing (ENG)
Rankine Road (Taylor and Francis). – Basingstoke 5 (1984) 2. – p. 253–515
AS 12.29.05
Key words: Remote sensing

Various papers about sensing methods and equipment. Analysis of results of various types of survey such as vegetation and geological. Generally of peripheral interest to geodesists and land surveyors.
Cotter, National Correspondent, of Great Britain IGD 84/III 115

AIDOS Systematics · **12.29.09** **Group III Photogrammetry and remote sensing**

Terrestrial photogrammetry

Methods of terrestrial survey
Use of instruments in terrestrial photogrammetry

Gruppe III Photogrammetrie und Fernerkundung

Terrestrisch-photogrammetrische Aufnahme

Verfahren der terrestrisch-photogrammetrischen Aufnahme
Nutzung von Geräten der terrestrisch-photogrammetrischen Aufnahme

528.71(21-194.2):528.711.11

Nemeth, G.
A lokalis fotogrammetria jelentoesege es a megvalositas lehetoesegei
Geod. es Kartogr. – Budapest 35 (1983) 4. – p. 274–278 : 4 fig. 7 lit.
The importance of local photogrammetry and the possibilities of its realization (HUN)
AS 12.29.09
Key words: Aerotriangulation, local photogrammetry

The article first reviews briefly some of the possibilities for the use of local photogrammetry, the camera carriers and the requirements to be met by aerotriangulation camera. An experiment intended to study, whether the kite as an unexpensive and for everybody accessible-tool of local photogrammetry is a suitable carrier for the purposes of aerotriangulation covering smaller areas, is set forth. The study tries also to answer the question, whether the imagery contains sufficient information to solve interpretation and metric tasks.

Veroe-Hetenyi, National Correspondent, Hungary IGD 84/III 116

528.711.11:528.74:72

Stambouloglou, E.
The elevator-camera system (ENG)
In: Precision and speed in close range photogrammetry : ISPRS Sympos., York, England, 5th to 10th September, 1982. — London, 1982. — (Internat. Archives Photogramm. 24 (1982) V/2). — p. 477—483 : 8 fig. 2 lit.
AS 12.29.09
Key words: Architectural photogrammetry, horizontal photography

A simple in construction, easy in handling, and economic system for horizontal photography, from heights up to 9 meters is described. Examples of applications in architectural photogrammetry are also presented.

(Author's summ.)

Techn. Univ. Dresden IGD 84/III 117

528.711.114:535.317.25

Wuertz, Gerhard
Abbildungseigenschaften der UMK 6,5/1318
Vermessungstechnik. — Berlin 32 (1984) 3. — p. 90—91 : 6 fig. 1 tab. 4 lit.
Representation properties of the UMK 6.5/1318 (GER)
AS 12.29.09/73.11
Key words: Effenciency parameter, terrestrial camera

Description of the efficiency of the UMK 6.5/1318 with respect to light distribution in the image plane, resolution power in the form of AWAR-values, and radial distortion. Detailed image representations give some information on the dependence of efficiency parameters on the choice of object distance, aperture adjustment, and on the film/filter combination. For this purpose a survey is given on the working fields to obtain best AWAR-values.

(Author's summ.)

Techn. Univ. Dresden IGD 84/III 118

AIDOS Systematics **12.29.13** **Group III Photogrammetry and remote sensing**

Photogrammetric acquisition of information from the air

Using of equipment and techniques of air photography and of aerocosmic survey
Using of accessories
Survey aircraft and survey flight operations
Survey flight planning
Aerial photography

Gruppe III Photogrammetrie und Fernerkundung

Photographische Erfassung von Informationen aus der Luft

Nutzung von Ausrüstungen und Verfahren der Luftbild- und aerokosmischen Aufnahme
Nutzung von Hilfsgeräten
Bildflugzeuge und Bildflugarbeiten
Bildflugplanung
Aerophotographie

528.711.1:629.733
528.77:930.26

Heckes, J.
Der Heissluftfesselballon des Deutschen Bergbau-Museums
In: Precision and speed in close range photogrammetry : ISPRS Sympos., York, England, 5th to 10th September, 1982. — London, 1982. — (Internat. Archives Photogramm. 24 (1982) V/1). — p. 245—252 : 9 fig. 4 lit.
The tethered hot-air balloon of the German Mining Museum (GER)
AS 12.29.13
Key words: Archaeology, close-range photogrammetry

A tethered hot-air balloon is introduced. The use of this balloon for close-range aerial photography has proven it to be a robust, relatively easy-to-manage camera vehicle with adequate loading capacity. One result of the evaluation of stereoscopic aerial photograms taken with the Rollei SLX "partially calibrated" camera system is described.

(Author's summ.)

Techn. Univ. Dresden IGD 84/III 119

528.711.1.089.6(21-194.2)
528.74:930.26:629.735.4

Mauelshagen, L. / Przybilla, Heinz-Juergen / Thies, M.
Ein Beispiel fuer „Mixed Ranges"-Kalibrierung in der Nahbereichsbildmessung
In: Precision and speed in close range photogrammetry : ISPRS Sympos., York, England, 5th to 10th September, 1982. — London, 1982. — (Internat. Archives Photogramm. 24 (1982) V/2). — p. 319—326 : 7 fig. 13 lit.
An example for calibration by the method of "mixed ranges" in close-range photogrammetry (GER)
AS 12.29.13/12.29.25
Key words: Archaeology, close-range photogrammetry, helicopter

A few days before its destruction a part of the roman wall of the city of Cologne/ Federal Republic of Germany has been documented by means of minihelicopter-photogrammetry. The calculation of calibration-triangulation follows the method of "Mixed Ranges".

(Author's summ.)

Techn. Univ. Dresden IGD 84/III 120

528.715(21-194.2):527.6

Przybilla, Heinz-Juergen
Navigationssysteme fuer den Einsatz in der Nahbereichsluftbildphotographie
In: Precision and speed in close range photogrammetry : ISPRS Sympos., York, England, 5th to 10th September, 1982. — London, 1982. — (Internat. Archives Photogramm. 24 (1982) V/2). — p. 397—404 : 8 fig. 9 lit.
Navigation systems for application in low altitude aerial photography (GER)
AS 12.29.13/12.29.25
Key words: Aerophotogrammetry, close-range photogrammetry, navigation

Constructions of navigation systems for low altitude aerial photography are discussed. The efficiency of the systems is shown; practical examples are given.

(Author's summ.)

Techn. Univ. Dresden IGD 84/III 121

528.711.1:629.733

Robinson, Major G P G
A helium balloon system for low altitude non-metric vertical photography (ENG)
In: Precision and speed in close range photogrammetry : ISPRS Sympos., York, England, 5th to 10th September, 1982. — London, 1982. — (Internat. Archives Photogramm. 24 (1982) V/2). — p. 412—421 : 3 fig. 2 tab. 4 lit.
AS 12.29.13/12.29.25
Key words: Air photography, balloon photogrammetry

In 1978 the author was tasked by The Directorate of Military Survey with examining the potential for employing a tethered balloon as a camera platform for survey control spotting photography. Various options and equipments were considered, a limited trial conducted and costed, and recommendations made for a mobile system. Some additional applications were also assessed.

(Author's summ.)

Techn. Univ. Dresden IGD 84/III 122

528.74:629.78

Wijk, M. C. van / Kratky, V.
Photogrammetric testing of the Space Shuttle remote manipulator performance (ENG)
In: Precision and speed in close range photogrammetry : ISPRS Sympos., York, England, 5th to 10th September, 1982. — London, 1982. — (Internat. Archives Photogramm. 24 (1982) V/2). — p. 551—557 : 4 fig.
AS 12.29.13
Key words: Calibration, movie camera, Space Shuttle

Six 16 mm movie cameras, installed in the payload bay of the Space Shuttle "Columbia", are used to monitor the performance of the remote manipulator system during several flight missions. Calibration procedures carried out in the laboratory and on board of the Space Shuttle are described. The accuracy of the photogrammetrically compiled information and initial results are discussed.

(Author's summ.)

Techn. Univ. Dresden IGD 84/III 123

061.3(100+713)"1982"ISPRS:528.7/.8

. . .
Proceedings — Commission II Symposium on "Advances in instrumentation for processing and analysis of photogrammetric and remotely sensed data" (ENG)
ISPRS — Ottawa, Canada, August 30 — September 3, 1982.
Internat. Archives of Photogramm. — Ottawa (1982) 24-II. — 704 p. : div. fig. div. tab. div. lit.
AS 12.29.13/12.29.41/12.01.17/73.11
Key words: Instruments of remote sensing, photogrammetric instruments

Working groups of Commission II

1. Analytical and hybrid photogrammetric instruments (p. 1—116)
2. Automated photogrammetric instruments and systems (p. 117—212)
3. Instruments for analysis of remotely sensed data (p. 213—331)
4. Instruments for preprocessing, storage and dissemination of remotely sensed data (p. 333—429)
5. Equipment for processing synthetic aperture radar data (p. 431—590)
6. Manufacturers' forum (p. 591—697).

Techn. Univ. Dresden IGD 84/III 124

AIDOS Systematics **12.29.21** **Group III Photogrammetry and remote sensing**

Photogrammetric determination of control points

Selection of control points for the photogrammetric and geodetic determination
Photogrammetric work for determining control points (as junction points for geodetic measurements)
Radial triangulation, aerial triangulation
Analytical aerotriangulation
Adjustment of strip and block triangulation

Gruppe III Photogrammetrie und Fernerkundung

Photogrammetrische Bestimmung von Paßpunkten

Auswahl von Paßpunkten für die photogrammetrische und geodätische Bestimmung
Photogrammetrische Arbeiten zur Bestimmung von Festpunkten (als Anschlußpunkte für geodätische Messungen)
Radialtriangulation, Aerotriangulation, analytische Aerotriangulation
Ausgleichung von Streifen- und Blocktriangulation

528.7(21-194.2):528.737.2

Abdel-Aziz, Youssef I.
The accuracy of control points for close range photogrammetry (ENG)
In: Precision and speed in close range photogrammetry : ISPRS Sympos., York, England, 5th to 10th September, 1982. — London, 1982. — (Internat. Archives Photogramm. 24 (1982) V/1). — p. 1—11 : 2 fig. 1 tab. 5 lit.
AS 12.29.21/12.29.33
Key words: Close range photogrammetry, control point

Control points are needed in close-range photogrammetry for orientation of the photographed object and for camera calibration. The intersection method by two theodolites is used for providing the photographed object with the necessary control points. The accuracy of these control points is very important in estimating the accuracy of the photogrammetric measurements.

This article gives the mathematical and the experimental proofs of newly developed formulas for estimating the accuracy of control point coordinates for any theodolite positions and the optimum theodolite positions that maximize the accuracy of these control points.
(Author's summ.)

Techn. Univ. Dresden IGD 84/III 125

528.735.2:528.16.088

Parsic, Z.
Untersuchungen ueber die Genauigkeit der Blockausgleichung nach der Buendelmethode
Vermess. Photogramm. Kulturtechnik. — Zuerich 81 (1983) 10. — p. 359—368 : 5 fig. 11 tab. 16 lit.
Investigations on the accuracy of block adjustment by the bundle method (GER)
AS 12.29.21/12.15.37
Key words: Block adjustment, bundle method

Three blocks with different material of image points are studied to evaluate the accuracy of block adjustment by the bundle method.

The investigation results showed that the comparison between the effective results (block 1) and results of simulated data (block 3) provides no satisfactory agreement. Hence, the following problem arises, how to evaluated by present means the efficiency from available information.

(Author's summ.)

Techn. Univ. Dresden IGD 84/III 126

AIDOS Systematics **12.29.25 Group III Photogrammetry and remote sensing**

Photogrammetric plotting

Fundamentals of the photogrammetric single-photograph plotting and double-image plotting
Graphical, optical-mechanical and analytical rectification, differential rectification, rectification with subjective and mechanical projection
Working with stereoplotting instruments (orientation)
Analytical plotting
Using of photogrammetric plotting instruments (comparators, analogue plotters, analytical plotting, stereoplotters, differential rectifiers, point transfer instruments)
Automation of the plotting processes
Digital recording of measurement results

Gruppe III Photogrammetrie und Fernerkundung

Photogrammetrische Auswertung

Grundlagen der photogrammetrischen Einzelbild- und Zweibildauswertung
Graphische, optisch-mechanische, analytische Entzerrung, Differentialentzerrung, Entzerrung mit subjektiver und mechanischer Projektion
Arbeit an Stereoauswertegeräten (Orientierung)
Analytische Auswertung
Nutzung von photogrammetrischen Auswertegeräten (Komparatoren, Analogauswertegeräte, analytische Auswertegeräte, Stereoauswertegeräte, Differentialentzerrungsgeräte, Punktübertragungsgeräte)
Automatisierung der Auswertungsprozesse
Digitale Erfassung der Meßergebnisse

528.721.2:519.213.3

Claus, Michael
Korrelationsrechnung in Stereobildpaaren zur automatischen Gewinnung von digitalen Gelaendemodellen, Orthophotos und Hoehenlinienplaenen
Karlsruhe, Univ., Fak. Bauing.- u. Vermess.-Wesen, Diss.
Dt. geod. Kommiss., R. C. — Muenchen (1983) 283. — 96 p. : 61 fig. 79 lit.
Computation of correlation in stereo pairs for the automatic aquisition of digital terrain models, orthophotos and maps of contour lines (GER)
AS 12.29.25/12.29.29
Key words: Computation of correlations, Orthophoto, Stereophotography

The present article introduces a technique which produces fully automatically terrain models and their products as orthophotos and contour line maps, from a digitzed stereo pair with a format of 23×23 cm^2. By means of the present programme version, terrain models of lower point density (30,000 points) may be calculated on a minicomputer (PRIME 500) by one-dimensional correlation along epipolar rays.

(Author's summ., abbr.)

Techn. Univ. Dresden IGD 84/III 127

528.854.4

Blum, Edgar
Ein interaktives Verfahren zur Klassifizierung von multivariaten Fernerkundungs-Bilddaten
Abhandl. z. Erlang. d. Titels eines Doktors d. Naturwiss. d. ETH Zuerich
Zuerich: Diss. ETH (1982) 7109. 115 p. : div. fig. div. tab. 54 lit.
An interactive procedure for the classification of multivariate remote sensing image data (GER)
AS 12.29.25/12.29.41
Key words: Application to multitemporal Landsat data, data by airborne Bendix M2S Scanner, interactive classification of remote sensing image data, parallelepiped

Description of an interactive procedure for the classification of remote sensing image data. Flexibility is guaranteed by a modular structure of the software in the form of individual functions. The classification routine is principally based on the classical parallelepiped method which uses class boundaries that are parallel to the co-ordinate axes. The classification parameters are defined interactively. The method is applied to a multitemporal set of Landsat data and multispectral data collected by an airborne Bendix M2S Scanner.

(Author's summ., abbr.)

Techn. Univ. Dresden IGD 84/III 128

528.7(21-194.2):681.3
528.722.6

Fuchs, H. / Leberl, F.
CRISP. A software package for close range photogrammetry for the Kern DSR-1 analytical stereoplotter (ENG)
In: Precision and speed in close range photogrammetry : ISPRS Sympos., York, England, 5th to 10th September, 1982. – London, 1982. – (Internat. Archives Photogramm. 24 (1982) V/1). – p. 175–184 : 2 fig. 4 lit.
AS 12.29.25/63.23
Key words: Analytical photogrammetry, close range photogrammetry stereoplotter

A software-package for close-range photogrammetry for the analytical stereoplotter DSR-1 of Kern & Co. Ltd., Aarau is being developed. The program-system serves to process metric- and non-metric camera photography, which may exist in the form of single images, single stereo-models and image-blocks. In addition to common control point coordinates the system will use known terrestrial observations such as distances, angles etc.

(Author's summ.)

Techn. Univ. Dresden IGD 84/III 129

528.711.114
528.74:72

Kis Papp, Laszlo
Primenenie fotogrammetricheskogo oborudovanija proizvodstva Cejjss-Ena dpja reshenija stroitel'nykh i arkhitekturnykh zadach
Period. Polytechn. : Ser. Civil Engin. – Budapest 27 (1983) 3–4. – p. 149–156 : 6 fig. 2 tab. 6 lit.
Photogrammetric equipment made by Carl Zeiss Jena for civil engineering and architectural applications (RUS)
AS 12.29.25/12.29.33/73.11
Key words: Architectural engineering, civil engineering, photogrammetric equipment

Carl Zeiss Jena have developed a photogrammetric camera and processing system for architectural and civil engineering applications.

Description is given of how to make photographs by use of SMK and UMK cameras, indicating photograph scales and explaining how to prepare design drawings.
Also the stereo-photogrammetric apparatus Technokart is presented.

Veroe-Hetenyi, National Correspondent, Hungary IGD 84/III 130

528.735:528.14:681.3

Larsson, Rune
Gentri — a system for simultaneous adjustment of photogrammetric and other observations (ENG)
In: Precision and speed in close range photogrammetry : ISPRS Sympos., York, England, 5th to 10th September, 1982. — London, 1982. — (Internat. Archives Photogramm. 24 (1982) V/2). — p. 301—310 : 1 fig. 15 lit.
AS 12.29.25/63.23
Key words: Bundle block adjustment

Report of the program and data structures as realized in the Gentri system. Implementation and results from some tests using real data. Further modifications by adding new types of information is discussed.

(Author's summ.)

Techn. Univ. Dresden IGD 84/III 131

528.721.2(21-194.2):528.14

Shortis, Mark R.
Sequential adjustments of close range stereopairs (ENG)
In: Precision and speed in close range photogrammetry : ISPRS Sympos., York, England, 5th to 10th September, 1982. — London, 1982. — (Internat. Archives Photogramm. 24 (1982) V/2). — p. 461—470 : 3 fig. 1 tab. 6 lit.
AS 12.29.25
Key words: Close range photogrammetry, sequential adjustment

Online analytical solutions for close range stereopairs are generally carried out using independent interior and exterior orientations with fixed control. Sequential adjustment allows the integration of all the control survey measurements and photogrammetric observations, significantly improving the restitution of the stereopair while avoiding the problems associated with handling large masses of data.

(Author's summ.)

Techn. Univ. Dresden IGD 84/III 132

528.721.126.088:681.3

Ugrin, N.
Genauigkeitsanalyse der orthophotoskopischen Verfahren
Period. Polytechn. : Ser. Civil Engin. — Budapest 27 (1983) 3—4. — p. 227—238 : 5 fig. 6 tab. 5 lit.
Precision analysis of orthophotoscopic methods (GER)
AS 12.29.25
Key words: Digital cartography, errors in orthophotoscopy, orthophotoscopic method

Development of orthophotoscopic instruments has much improved their economy and precision. Errors encountered in applying the method will be pointed out, and so will be the sources of error, giving hints of how to avoid them.

Veroe-Hetenyi, National Correspondent, Hungary IGD 84/III 133

528.72

Ugrin, N.
Kepatalakitas eloere szamitott adatokkal
Geod. es Kartogr. – Budapest 35 (1983) 4. – p. 266–271 : 7 fig. 1 tab. 11 lit.
Transformation of images using precomputed data (HUN)
AS 12.29.25
Key words: Exterior orientation, image transformation, optical transformation

Theory of optical transformation of images using precomputed data was known already in the '30s but it has not been spread in practice so far. Its reasons lie in the expensiveness of calculation without a programmable computer, in lack of a correct method for measuring or computing input data as well as in the frequent misinterpretation of input data.

The computer solution of the three point problem in space permits the reconsideration of the problem and the development of the method.

Novelty of this method is to take input data from results of the three point problem in space.

It is unambiguously justified by experiments, the accuracy of the method of analytical transformation of images is equal to that of transformation of images by points of minor control.

A forecast is given on possibilities of instrument development in chapter 5.

Veroe-Hetenyi, National Correspondent, Hungary IGD 84/III 134

528.721.1.089.6

Wester-Ebbinghaus, Wilfried
Einzelstandpunkt-Selbstkalibrierung – mathematische Formulierung und erste Erfahrungen
In: Precision and speed in close range photogrammetry : ISPRS Sympos., York, England, 5th to 10th September, 1982. – London, 1982. – (Internat. Archives Photogramm. 24 (1982) V/2). – p. 533–550 : 15 fig. 3 tab. 14 lit.
Single station self calibration – mathematical formulation and first experiences (GER)
AS 12.29.25
Key words: Camera calibration, single station self calibration

This new procedure allows a full camera calibration by means of perspective restrictions, using photos with different orientation angles from one single station. There is no need of any geodetic control.

(Author's summ.)

Techn. Univ. Dresden IGD 84/III 135

061.3.053.52(430-2.6)"1982"
681.2.056:528.854

. . .
Proceeding 6th International Conference on Pattern Recognition, Munich, FRG, Oct. 19–22, 1982
Dt. Arbeitsgemeinschaft Mustererkennung. – Silver Spring (USA) : IEEE Computer Soc. Press, 1982. – Vol. 1 and 2. – 568 and 670 p. : div. fig. div. tab. div. lit.
AS 12.29.25/12.29.37
Key words: Pattern recognition, photogrammetry, remote sensing, symposium 1982

Techn. Univ. Dresden IGD 84/III 136

AIDOS Systematics **12.29.29** **Group III Photogrammetry and remote sensing**

Application of photogrammetric techniques to map production

Production and revision of topographic and thematic maps on all scales by means of aerial- and terrestrial-photogrammetric methods
Using of aerocosmic photographs to produce and revise maps

Gruppe III Photogrammetrie und Fernerkundung

Anwendung photogrammetrischer Verfahren zur Kartenherstellung

Herstellung und Laufendhaltung topographischer und thematischer Karten aller Maßstäbe mit aero- und terrestrisch-photogrammetrischen Verfahren
Nutzung aerokosmischer Aufnahmen zur Herstellung und Laufendhaltung von Karten

528.71:528.47

Fryer, J. G.
A simple system for photogrammetric mapping in shallow water (ENG)
Photogramm. Record. — London 11 (1983) 62. — p. 203—208 : 2 fig. 14 lit.
AS 12.29.29/12.29.25
Key words: Marine geodesy, stereophotography, underwater photography

A photogrammetric system that is suitable for shallow water detailed mapping at scales from 1 : 50 to 1 : 500 for coastal and harbour engineering projects is described. The results from tests using automatic 35 mm Canon cameras on an elevated platform to determine the accuracy of planimetric and depth co-ordinates are presented. The economic benefits of stereophotography through shallow wather rather than underwater photography are discussed.

(Authors summ.)

Cotter, National Correspondent of UK Great Britain IGD 84/III 137

528.721.126(083.4)

Nittinger, Johannes
Zur Bedeutung von Orthophotokarten in Entwicklungslaendern
Z. Vermess.-Wesen. — Stuttgart 108 (1983) 8. — p. 331—339 : 4 fig. 5 lit.
The importance of orthophotomaps in developing countries (GER)
AS 12.29.29/12.29.25
Key words: Orthophotomap, planning

Orthophotomaps are very suitable planning instruments, for planning purposes and at different planning stages. On the other hand orthophotos can be used as a basis for cadastral maps and recording of land register data. Some examples demonstrate that by employing technical assistance from industrialized countries this method can be a very practicable and economical way for solving different problems in developing countries.

(Author's summ.)

Techn. Univ. Dresden IGD 84/III 138

AIDOS Systematics **12.29.33** **Group III Photogrammetry and remote sensing**

Application of photogrammetry for non-topographic purposes

Photogrammetric determination of the shape, size and position of non-topographic objects
Close-range photogrammetry, space-time photogrammetry
Application of unconventional imaging techniques in the close-range

Gruppe III Photogrammetrie und Fernerkundung

Anwendung der Photogrammetrie für nicht-topographische Zwecke

Photogrammetrische Form-, Größen- und Lagebestimmung nicht-topographischer Objekte
Nahphotogrammetrie, Raum-Zeit-Photogrammetrie
Anwendung nicht-konventioneller Abbildungsverfahren im Nahbereich

528.73(26.02):513.012

Adams, L. P.
Underwater analytical photogrammetry using non-metric cameras (ENG)
In: Precision and speed in close range photogrammetry : ISPRS Sympos., York, England, 5th to 10th September, 1982. — London, 1982. — (Internat. Archives Photogramm. 24 (1982) V/1). — p. 12—22 : div. fig. 4 lit.
AS 12.29.33
Key words: Analytical photogrammetry, underwater photogrammetry

A pair of NIKONOS III underwater cameras equipped with a variety of interchangeable lenses has been used to produce photography in air and underwater. The distortion characteristics of the various lenses have been studied and the camera has been tested when the resulting photography is employed in a stereophotogrammetric analytical mode. Accuracy comparisons are made using measurements taken on a precision comparator utilising the original negatives and on paper print positive enlargements observed in a stereoscopic mode in which the photographs are mounted on a digitiser table and viewed stereoscopically through a modified mirror stereoscope. A dual cursor is employed to provide a stereocomparator facility.

(Author's summ.)

Techn. Univ. Dresden IGD 84/III 139

528.74:624.19

Anderson, H. / Stevens, D.
Mono photographic tunnel profiling (ENG)
In: Precision and speed in close range photogrammetry : ISPRS Sympos., York, England, 5th to 10th September, 1982. — London, 1982. — (Internat. Archives Photogramm. 24 (1982) V/1). — p. 23—30 : 3 fig.
AS 12.29.33/12.29.09/83
Key words: Stereometric photogrammetry, tunnel profiling

When the search for an alternative to stereometric photogrammetry as a tunnel profiling tool was investigated we hoped to achieve a system that reduced field and laboratory time, thus reducing cost and improving accuracy. Results show we have succeeded on both counts.

(Author's summ.)

Techn. Univ. Dresden IGD 84/III 140

528.74:72

Atkinson, K. B.
Castle howard revisited (ENG)
In: Precision and speed in close range photogrammetry : ISPRS Sympos., York, England, 5th to 10th September, 1982. — London, 1982. — (Internat. Archives Photogramm. 24 (1982) V/1). — p. 31—38 : 6 fig.
AS 12.29.33
Key words: Architectural photogrammetry

In 1962, E. H. Thompson published the results of an unorthodox application of

photogrammetry to the solution of a major problem in the restoration of Castle Howard. This International Symposium provides an appropriate opportunity to recall a pioneering example of British architectural photogrammetry.
(Author's summ.)
Techn. Univ. Dresden IGD 84/III 141

528.71(26.02)

Baldwin, R. A. / Newton, I.
A proposed system for underwater photogrammetry from a manned submersible camera platform (ENG)
In: Precision and speed in close range photogrammetry : ISPRS Sympos., York, England, 5th to 10th September, 1982. — London, 1982. — (Internat. Archives Photogramm. 24 (1982) V/1). — p. 39—52 : 6 fig. 3 tab. div. lit.
AS 12.29.33
Key words: Close range photogrammetry, economy

Systems for use in low altitude aerial photography are presented, costs and accuracy of these systems are compared with conventional procedures. Plotting methods are discussed and explained by an example.
(Author's summ.)
Techn. Univ. Dresden IGD 84/III 142

528.71.003.1(21-194.2)

Beckschaefer, Martin / Przybilla, Heinz-Juergen
Wirtschaftlichkeitsfragen in der Nahbereichsluftbildmessung
In: Precision and speed in close range photogrammetry : ISPRS Sympos., York, England, 5th to 10th September, 1982. — London, 1982. — (Internat. Archives Photogramm. 24 (1982) V/1). — p. 53—61 : 6 tab. 14 lit.
Economy problems in close range photogrammetry (GER)
AS 12.29.33
Key words: Close range photogrammetry, economy

Systems for use in low altitude aerial photography are presented, costs and accuracy of these systems are compared with conventional procedures. Plotting methods are discussed and explained by an example.
(Author's summ.)
Techn. Univ. Dresden IGD 84/III 143

528.711.11(21-194.2):681.3

Brown, Duane C.
STARS (Simultaneous Triangulation And Resection System). A turnkey system for close range photogrammetry (ENG)
In: Precision and speed in close range photogrammetry : ISPRS Sympos., York, England, 5th to 10th September, 1982. — London, 1982. — (Internat. Archives Photogramm. 24 (1982) V/1). — p. 68—89 : 6 fig.
AS 12.29.33
Key words: Bundle adjustment, close range photogrammetry, computer, terrestrial camera

STARS is an all-inclusive system for close-range photogrammetry consisting of three major elements: (1) a new, large-format (23 × 23 cm) terrestrial camera with interchangeable lens cones, (2) a powerful, desktop computer system on which is installed (3) a bundle adjustment with self-calibration especially optimized for the diverse applications encountered in close-range photogrammetry. The system is designed to operate directly on the output of any of several commercially available monocomparators.
(Author's summ.)
Techn. Univ. Dresden IGD 84/III 144

528.74.088:72

Carbonnell, Maurice
Precision des releves photogrammetriques d'architecture
In: Precision and speed in close range photogrammetry : ISPRS Sympos., York, England, 5th to 10th September, 1982. — London, 1982. — (Internat. Archives Photogramm. 24 (1982) V/1). — p. 112—118
Precision in architectural photogrammetry (FRA)
AS 12.29.33
Key words: Architectural photogrammetry

The notions regarding the accuracy of an architectural photogrammetry survey are rather complex. It is necessary to consider the accuracy in all its aspects. On the one hand, the architectural aspects: reference planes, content of the survey, state of conservation of the building, influence of the operator's interpretation, degree of accuracy required. On the other hand, the technical aspects of the photogrammetric process in the particular case of the architectural surveys: camera set-ups in relation to the building, considerable depth of the subject, large obliquity of the camera axes, lighting conditions.

(Author's summ.)

Techn. Univ. Dresden IGD 84/III 145

528.74:72
528.722:681.3

Dallas, R. W. A.
Low cost digitisation of an analogue plotter (ENG)
In: Precision and speed in close range photogrammetry : ISPRS Sympos., York, England, 5th to 10th September, 1982. — London, 1982. — (Internat. Archives Photogramm. 24 (1982) V/1). — p. 119—128 : 7 fig.
AS 12.29.33/73.11
Key words: Architectural photogrammetry

It is frequently valuable in architectural photogrammetry to be able to provide interchangeable output between XYZ axes to facilitate the plotting of cross-sections etc. The paper describes the low cost installation of a microcomputer based system to an analogue plotter (Thompson-Watts Mark II). The system records X, Y, Z coordinates of the model using rotary incremental encoders and then drives the existing coordinatograph table through stepper motors.

(Author's summ.)

Techn. Univ. Dresden IGD 84/III 146

528.74:551.462

Fryer, John G.
Photogrammetry through shallow water (ENG)
Austral. J. Geod., Photogramm. and Survey. — Sydney (1983) 38. — p. 25—38•: 2 fig. 20 lit.
AS 12.29.33
Key words: Analytical stereoplotter, shallow water, underwater photography

The use of photography from a floating platform to aid underwater ecological studies on coral reef flats is proposed. Alternative solutions are investigated and a floating tetrahedron described. The difficulties of using an analytical stereoplotter with two-media photography are detailed and a summary presented of the relevant mathematics.

(Author's summ.)

Techn. Univ. Dresden IGD 84/III 147

528.72.088(21-194.2)

El-Beik, A. H. A. / Babaei-Mahani, R.
Quadrustational close range photogrammetric system (ENG)
In: Precision and speed in close range photogrammetry : ISPRS Sympos., York, England, 5th to 10th September, 1982. — London, 1982. — (Internat. Archives Photogramm. 24 (1982) V/1). — p. 129—139 : 9 fig. 3 tab. 5 lit.
AS 12.29.33
Key words: Close range photogrammetry

Mathematical models representing the geometry of multistation/multistereo as well as quadrustational systems were developed for the normal case and the general case of photogrammetry using optimisation from rays and planes. The accuracy of the system was assessed by taking stereo-comparator measurements from steropairs of a control field.
(Author's summ.)
Techn. Univ. Dresden IGD 84/III 148

528.74(21-194.2):624.9
528.721.2:681.3

Faig, W. / El-Hakim, S. F.
The use of distances as object space control in close range photogrammetry (ENG)
In: Precision and speed in close range photogrammetry : ISPRS Sympos., York, England, 5th to 10th September, 1982. — London, 1982. — (Internat. Archives Photogramm. 24 (1982) V/1). — p. 144—148
AS 12.29.33/12.29.25
Key words: Bundle adjustment, close range photogrammetry

GEBAT-V, the close-range version of a bundle program with self calibration for simultaneous adjustment of photo coordinates, object distances and height differences is briefly discussed. Test results of an industrial project using distances in various combinations are presented.
(Author's summ.)
Techn. Univ. Dresden IGD 84/III 149

528.74:(528.711.11)(26.02)(94)

Fryer, John G.
Underwater 35 mm photogrammetric applications in Australia (ENG)
In: Precision and speed in close range photogrammetry : ISPRS Sympos., York, England, 5th to 10th September, 1982. — London, 1982. — (Internat. Archives Photogramm. 24 (1982) V/1). — p. 167—174 : 4 fig.
AS 12.29.33/73.11
Key words: Underwater camera, underwater photogrammetry

Applications of non-metric 35 mm underwater cameras to obtain data for an off-shore pipeline project and a long term study of coral ecology are presented. The former project uses an image splitting attachment to obtain stereoscopic half-frames whilst the latter uses a twin camera rig. The analytical techniques used to process the data are discussed.
(Author's summ.)
Techn. Univ. Dresden IGD 84/III 150

528.7.088(21-194.2)

Gates, J. W. C. et al.
Factors defining precision in close range photogrammetry (ENG)
In: Precision and speed in close range photogrammetry : ISPRS Sympos., York, England, 5th to 10th September, 1982. — London, 1982. — (Internat. Archives Photogramm. 24 (1982) V/1). — p. 185—195 : 5 fig. 5 lit.
AS 12.29.33/12.29.25
Key words: Close range photogrammetry, precision

This paper describes an investigation of the factors limiting the precision possible with conventional close range photogrammetry.
Several problem areas have been identified, and possible solutions are suggested; particular study has been given to the form and environmental stability of photographic plates. Two photogrammetric tests are also described. The first is a "conventional" survey of a cubic space frame, using mult-station recording, and the second is a survey of a threeaxis measuring machine using two-camera convergent photogrammetry, the machine being used as a space frame.
(Author's summ.)
Techn. Univ. Dresden IGD 84/III 151

528.711:629.73

Georgopoulos, A.
Low altitude non-metric photography using a kite (ENG)
In: Precision and speed in close range photogrammetry : ISPRS Sympos., York, England, 5th to 10th September, 1982. — London, 1982. — (Internat. Archives Photogramm. 24 (1982) V/1). — p. 207—217 : 1 fig. 5 tab. 9 lit.
AS 12.29.33
Key words: Kite photography, non-metric photography

A kite photography system is described, which involves non-metric photography and the use of simple instrumentation for processing the data. A practical test is described and the results are assessed for their accuracy, usefulness and cost effectiveness.
(Author's summ.)
Techn. Univ. Dresden IGD 84/III 152

528.74:61

Gibaud, B. et al.
Accurate use of CT. scanner, stereotaxis and photogrammetry for diagnosis and treatment of cerebral lesions (Tumours) (ENG)
In: Precision and speed in close range photogrammetry : ISPRS Sympos., York, England, 5th to 10th September, 1982. — London, 1982. — (Internat. Archives Photogramm. 24 (1982) V/1). — p. 225—234 : 7 fig. 9 lit.
AS 12.29.33
Key words: Computed tomography, head fixation system, stereotaxis

The authors explain the present state of their research on an accurate system for analysis and combination for three-dimensional information from C. T. Scanner, stereotaxis and stereoscopy.
The basic problem is the transfer of co-ordinates between these different investigation techniques. The theoretically simple geometrical problem has proven quite difficult in practice. A careful study of a reference device is needed. A second important point is the organization of the system which must not ignore the technical, medical and economical contexts. This system consists in the computing unit of an analytical stereoplotter, connected to peripheral units that will be further detailed.
(Author's summ.)
Techn. Univ. Dresden IGD 84/III 153

528.74(21-194.2)

Granshaw, S. I.
Precision and the multistation bundle method: Concepts and preconceptions (ENG)
In: Precision and speed in close range photogrammetry : ISPRS Sympos., York, England, 5th to 10th September, 1982. — London, 1982. — (Internat. Archives Photogramm. 24 (1982) V/1). — p. 235—244 : 3 fig. 1 tab. div. lit.
AS 12.29.33
Key words: Bundle method, close range photogrammetry

The co-ordination of targetted close range structures using the multistation bundle method provides an opportunity to clarify concepts of precision and accuracy, and their means of attainment, freed from restrictive preconceptions fostered by stereoscopic photogrammetry.

(Author's summ.)

Techn. Univ. Dresden IGD 84/III 154

528.74:629.12

Holm, Knut Ragnar / Oestbye, Bjoern
The FOMAKON project, photogrammetry on marine structures (ENG)
In: Precision and speed in close range photogrammetry : ISPRS Sympos., York, England, 5th to 10th September, 1982. – London, 1982. – (Internat. Archives Photogramm. 24 (1982) V/1). – p. 253–260 : 3 fig. 7 lit.
AS 12.29.33
Key words: Photogrammetry on marine structure, shipyard

At shipyards dimensional checking by use of photogrammetry is most often carried out as ad hoc solutions. The aim of the project described is to develop photogrammetric procedures well suited for the specific demands and problems met at shipyards, and to make this an integrated part of the yard's Quality Control Program.

(Author's summ.)

Techn. Univ. Dresden IGD 84/III 155

528.74:61

Karras, G. E. / Petsa, E. P. / Tympanidis, K. N.
Lower limb measurements using moire photography (ENG)
In: Precision and speed in close range photogrammetry : ISPRS Sympos., York, England, 5th to 10th September, 1982. – London, 1982. – (Internat. Archives Photogramm. 24 (1982) V/1). – p. 261–271 : 4 fig. 3 tab. div. lit.
AS 12.29.33
Key words: Anthropometry, moire photography, photogrammetry

Using truncated cone formulae for legs (gluteal furrow to maximum calf) partitioned into five segments, leg volume and surface area values for a sample of 36 subjects were determined from both direct anthropometric and simple (diameter) measurements on moire photographs. Comparison of the two approaches showed that leg circumferences can be satisfactorily approximated by similar sections; and that volume and surface area differences do not exceed 4 % (mean differences are practically zero).

(Author's summ.)

Techn. Univ. Dresden IGD 84/III 156

528.711.11.089.6:528.73
528.74:72

Kleinmann, R. / Mauelshagen, L. / Wester-Ebbinghaus, W.
Simultankalibrierung einer Teil-Messkammer unter Verwendung von Zusatzbeobachtungen im Objektraum
In: Precision and speed in close range photogrammetry : ISPRS Sympos., York, England, 5th to 10th September, 1982. – London, 1982. – (Internat. Archives Photogramm. 24 (1982) V/1). – p. 272–281 : 10 fig. 3 tab. 7 lit.
Simultaneous calibration of a measuring camera using additional observations in object space (GER)
AS 12.29.33/65.33
Key words: Architecture photogrammetry, calibration, control point

The actual situation of the mosaic floor in the Kapitel-hall of the monastery Rommersdorf near Neuwied/Fed. Rep. of Germany had been documented by photogrammetry for restoration works. By means of simultaneous calibration (bundle triangulation) the efficiency of Rolleiflex SLX will be tested for determination of control points.

(Author's summ.)

Techn. Univ. Dresden IGD 84/III 157

528.737:72

Kotowski, Ruediger / Roessmann, Heiner / Wester-Ebbinghaus, Wilfried
Zweischalige Buendeltriangulation an einem Grossbauwerk
In: Precision and speed in close range photogrammetry : ISPRS Sympos., York, England, 5th to 10th September, 1982. — London, 1982. — (Internat. Archives Photogramm. 24 (1982) V/2). — p. 282—291 : 12 fig. 6 lit.
Two-sheet bundle triangulation on a large building (GER)
AS 12.29.33/83
Key words: Architecture, point determination

A procedure for simultaneous photogrammetric point determination inside and outside a large building is presented shown at the church San Domenico in Siena/Italy. Characteristics: Use of the semi-metric camera (Teil-Messkammer) Rolleiflex SLX, high flexibility in the choice of suitable camera stations, use of distances and elevation differences as the only geodetic control and interactive data reduction.

(Author's summ.)

Techn. Univ. Dresden IGD 84/III 158

528.7(21-194.2)

Kupfer, Guenther
Aktuelle Fragen zur Strategie in der Nahbereichsphotogrammetrie
In: Precision and speed in close range photogrammetry : ISPRS Sympos., York, England, 5th to 10th September, 1982. — London, 1982. — (Internat. Archives Photogramm. 24 (1982) V/2). — p. 292—300 : 1 fig. 1 tab.
Topics concerning the strategy in close-range photogrammetry (GER)
AS 12.29.33
Key words: Close-range photogrammetry

Today photogrammetric end products may be achieved by manyfold means. Thus thorough strategical planning is essential. Possible process components are listed, emphasizing actual developments.

(Author's summ.)

Techn. Univ. Dresden IGD 84/III 159

528.74(21-194.2):621

Monczkowski, Uwe / Regensburger, Karl
Photogrammetrische Prueftechnologie fuer Industrieroboter
Jenaer Rdsch. — Berlin 29 (1984) 1. — p. 28—31 : 5 fig. 1 tab. 4 lit.
Photogrammetric checking technology for industrial robots (GER)
AS 12.29.33/93
Key words: Industrial robot, photogrammetry

(s. a. IGD 84/III 39)

Techn. Univ. Dresden IGD 84/III 160

528.74:726.7(5)

Meid, Alfons / Hansch, Kay-Uwe
Photogrammetrische Aufnahme von Tempel- und Klosteranlagen in Kandy, Sri Lanka
Allg. Vermess.-Nachr. – Karlsruhe 91 (1984) 1. – p. 15–23 : 2 fig. 2 tab. 5 lit.
Photogrammetric survey of temple and cloister structures in Kandy, Sri Lanka (GER)
AS 12.29.33
Key words: Architectural photogrammetry
Techn. Univ. Dresden IGD 84/III 161

528.721.1
528.74:624

Miloshev, Stojancho
Kontrolirane na elementite na orientirane pri izledvane na deformacii na inzhenerni sahorahzhenija po fotogrametrichen nachin
Geod., Kartogr., Zemeustrojjstvo. – Sofija (1983) 5. – p. 32–36 : 2 fig. 7 tab. 2 lit.
The checking of orientation elements when studying deformation of engineering structures by photogrammetric method (BUL)
AS 12.29.33/12.29.25
Key words: Civil engineering, control point, orientation of photograms

A combination of two known methods is suggested for checking and taking into consideration the changes in elements of external orientation, by means of doubling geodetic measurements on appropriately located control points, when applying stereophotogrammetry in studying deformation of engineering structures. Results of a real experiment, according to the three methods mentioned, are given.
Georgiev, National Correspondent, Bulgarian IGD 84/III 162

528.74:61

Newton, I. / Craxford, A. D.
A photogrammetric method for measuring footprints in connection with foot disorders (ENG)
In: Precision and speed in close range photogrammetry : ISPRS Sympos., York, England, 5th to 10th September, 1982. – London, 1982. – (Internat. Archives Photogramm. 24 (1982) V/2). – p. 327–334 : 7 fig. 1 tab. 2 lit.
AS 12.29.33/91
Key words: Application of photogrammetry, medicine

An application of photogrammetry for measuring footprints is described. The footprints of the static foot in load are produced using expanded polyethylene foam (Plastazote) and are then measured photogrammetrically to provide a quantitative analysis of the sole of the foot. This analysis is applicable to clinical use in the treatment of foot disorders.
(Author's summ.)
Techn. Univ. Dresden IGD 84/III 163

528.7(21-194.2):513

Okamoto, Atsushi
Geometrical problem in close range photogrammetry (ENG)
In: Precision and speed in close range photogrammetry : ISPRS Sympos., York, England, 5th to 10th September, 1982. – London, 1982. – (Internat. Archives Photogramm. 24 (1982) V/2. – p. 335–347 : 1 fig. 22 lit.
AS 12.29.33/43.21
Key words: Close range photogrammetry, geometry, orientation parameter

This paper describes the simultaneous determination of all orientation parameters of overlapped photographs both for the general case where a picture has 11 independent central-projective elements and for a special case where the geometry of a photograph can be determined with nine independent central-projective parameters (six exterior and three interior orientation elements in conventional photogrammetry). Non-central-projective elements are considered separately, which describes the non-linear part of the disturbing features of non-metric cameras. In addition, another approach to the simultaneous determination is presented by adopting directly the collinearity equations in algebraic form as the determination equation. Object space control is established not only in terms of co-ordinates of points but also in terms of distances. To perform the practical test of the methods proposed is my next program.

(Author's summ.)

Techn. Univ. Dresden IGD 84/III 164

528.74:61

Petsa, E. P. / Karras, G. E. / Badekas, J.
Assessing the accuracy of moire topography measurements (ENG)
In: Precision and speed in close range photogrammetry : ISPRS Sympos., York, England, 5th to 10th September, 1982. — London, 1982. — (Internat. Archives Photogramm. 24 (1982) V/2). — p. 367—376 : 5 fig. 1 tab. 7 lit.
AS 12.29.33/12.29.25/91
Key words: Application of photogrammetry, medicine

The accuracy of shadow moire topography is investigated from the point of view of its applications in medicine. Known lengths, surface areas and volumes of seven three-dimensional objects were thus repeatedly determined from measurements on moire photographs. The resulting accuracy and precision of the method satisfy the requirements of most studies of body form.

(Author's summ.)

Techn. Univ. Dresden IGD 84/III 165

528.74(21-194.2):59

Ruether, H.
Wild life stereo photogrammetry at close range (ENG)
In: Precision and speed in close range photogrammetry : ISPRS Sympos., York, England, 5th to 10th September, 1982. — London, 1982. — (Internat. Archives Photogramm. 24 (1982) V/2). — p. 422—432 : 8 fig. 1 tab. 7 lit.
AS 12.29.33/91
Key words: Biology, close range photogrammetry

The close range photogrammetric survey of shoulder heights of African elephants in their natural habitat is described. Model deformations caused by errors in plate co-ordinates and elements of inner orientation are investigated. Simulated errors are introduced into computer generated stereo pairs and actual photography of a test field is studied. The efficiency of two simple stereoscopic model control methods ("covergency error correction" and "affine scaling") ist tested. Both methods rely on a few measured distances only. Equipment and field work is discussed. Results with standard deviation of ± 1 to 4 cm were obtained.

(Author's summ.)

Techn. Univ. Dresden IGD 84/III 166

528.721

Scott, P. J. / Stephens, C. D. / Butcher, G. W.
Applications of the reflex instruments (ENG)
In: Precision and speed in close range photogrammetry : ISPRS Sympos., York, England, 5th to 10th September, 1982. — London, 1982. — (Internat. Archives Photogramm. 24 (1982) V/2). — p. 451—460 : 3 fig. 5 lit.
AS 12.29.33/12.29.25/73.11
Key words: Reflex metrograph, reflex microscope

Several projects undertaken on the Reflex Metrograph and the Reflex Microscope are described. Evidence is given of their suitability, increased accuracy and greater flexibility compared with photogrammetry.
(Author's summ.)
Techn. Univ. Dresden IGD 84/III 167

528.74:551.32

Stirling, D. M.
Measuring short term glacial fluctuations by aerial and terrestrial photogrammetry – a comparative study (ENG)
In: Precision and speed in close range photogrammetry : ISPRS Sympos., York, England, 5th to 10th September, 1982. – London, 1982. – (Internat. Archives Photogramm. 24 (1982) V/2). – p. 484–496 : 7 fig. 2 tab. 10 lit.
AS 12.29.33/12.29.29
Key words: Aerophotogrammetry, glaciology, terrestrial photogrammetry
A comparison is made of the fieldwork requirements for mapping the Bas Glacier d'Arolla, Switzerland by terrestrial and aerial photogrammetric methods. Also included is a description of the development and testing of an on-line analytical solution for determining short term glacial fluctuations. The results from aerial and terrestrial photography are compared.
(Author's summ.)
Techn. Univ. Dresden IGD 84/III 168

528.74:629.12

Szczechowski, Bogdan
Photogrammetric study of the course of a ship launching (ENG)
In: Precision and speed in close range photogrammetry : ISPRS Sympos., York, England, 5th to 10th September, 1982. – London, 1982. – (Internat. Archives Photogramm. 24 (1982) V/2). – p. 497–506 : 3 fig. 3 lit.
AS 12.29.33
Key words: Application of photogrammetry, shipbuilding
I discuss the new possibilities of applications of photogrammetry to the studies of the behaviour of a ship body during its launching. As a result of appropriate works, simultaneous determination of the following parameters of a launched ship is made possible: location on the slip, heel, velocity, vertical and horizontal instantaneous deformation.
(Author's summ.)
Techn. Univ. Dresden IGD 84/III 169

528.711.1(26.02):622.323(261.2)

Turner, John / Leatherdale, John
Underwater photogrammetry for inspection and maintenance of North Sea oil platforms (ENG)
In: Precision and speed in close range photogrammetry : ISPRS Sympos., York, England, 5th to 10th September, 1982. – London, 1982. – (Internat. Archives Photogramm. 24 (1982) V/2). – p. 507–515 : 5 fig. 5 lit.
AS 12.29.33
Key words: Oil platform inspection, underwater photogrammetry
A commercial service for underwater photogrammetry has been operating since September 1980 providing cameras, expertise in underwater photography and photogrammetric measurements. This service is used routinely for the inspection and maintenance of oil production platforms in the North Sea and other applications are developing around the world. The cameras, underwater procedures and photogrammetric methods are described with examples of typical projects.
(Author's summ.)
Techn. Univ. Dresden IGD 84/III 170

528.74:624.97
528.718:681.3

Veress, S. A.
Photogrammetric analysis of deformation of transmission lines (ENG)
In: Precision and speed in close range photogrammetry : ISPRS Sympos., York, England, 5th to 10th September, 1982. – London, 1982. – (Internat. Archives Photogramm. 24 (1982) V/2). – p. 516–523 : 6 fig. 2 tab. 5 lit.
AS 12.29.33/12.29.25
Key words: Terrestrial photogrammetry, transmission line tower

Terrestrial photogrammetry was used to determine the structural deformation of an electric transmission tower and conductor under static and dynamic load. The photogrammetric methodology and the results of this project are given. The computer graphics are shown for rapid evaluation of the large amount of data obtained.

(Author's summ.)

Techn. Univ. Dresden IGD 84/III 171

528.74:62

Vyner, N. A. / Hanold, J. H.
A.I.M.S. – Analytical industrial measuring systems (ENG)
In: Precision and speed in close range photogrammetry : ISPRS Sympos., York, England, 5th to 10th September, 1982. – London, 1982. – (Internat. Archives Photogramm. 24 (1982) V/2). – p. 524–532
AS 12.29.33
Key words: Analytical measuring system, Keuffel and Esser analytical industrial measuring system

Techn. Univ. Dresden IGD 84/III 172

528.74(21-194.2):629.113

Wahl, M.
Photogrammetry at Regie Renault (ENG) *
Photogramm. Record. – London 11 (1983) 62. – p. 195–201 : 5 fig.
AS 12.29.33/12.29.25
Key words: Car accident investigation, close range photogrammetry

Photogrammetric techniques are of value in various fields of application, including the styling of new models of cars, design modifications to existing models, car accident investigations and statistical studies on human bodies.

* An account of this work was presented at the Internat. Sympos. on Precision and Speed in Close Range Photogrammetry, held at the Univ. York, Sept. 1982.

Cotter, National Correspondent of the UK Great Britain IGD 84/III 173

AIDOS Systematics **12.29.41** **Group III Photogrammetry and remote sensing**

Non-photographic recording of information from the air and its interpretation

Recording of information by non-photographic techniques and aerocosmic methods
Using of recording and imaging devices
Interpretation of non photographic images
Decoding of photoelectronic images
Application of the results of interpretation

Gruppe III Photogrammetrie und Fernerkundung

Nichtphotographische Erfassung von Informationen aus der Luft und ihre Interpretation

Erfassung von Informationen mit nichtphotographischen Verfahren und aerokosmischen Methoden
Einsatz von Erfassungs- und Aufnahmegeräten
Interpretation nichtphotographischer Aufnahmen
Dechiffrierung photoelektronischer Aufnahmen
Anwendung der Interpretationsergebnisse

528.711.7:528.8

Ghosh, Sanjib K.
Some ideas on measuring accuracy with electron micrography (ENG)
In: Precision and speed in close range photogrammetry : ISPRS Sympos., York, England, 5th to 10th September, 1982. — London, 1982. — (Internat. Archives Photogramm. 24 (1982) V/1). — p. 218—224 : 4 tab. 4 lit.
AS 12.29.41
Key words: Electron micrography system, stereotope

Accuracy of measurements with electron micrography seems to depend on factors like distortions, scale repeatability, intersection geometry, instrumentation, magnification, etc. Some ideas on these obtained from recent research investigations are presented with regard to one SEM system.

(Author's summ.)

Techn. Univ. Dresden IGD 84/III 174

AIDOS Systematics **73.11** **Group III Photogrammetry and remote sensing**

Design and checking of photogrammetric instruments

Devices, instruments and combined devices for terrestrial surveying
Instruments and combined devices for the photogrammetric plotting and interpretation (comparators, analogue plotting instruments, analytical plotters, stereoplotters, differential rectifiers, point transfer instruments)

Gruppe III Photogrammetrie und Fernerkundung

Konstruktion und Prüfung photogrammetrischer Geräte

Geräte, Instrumente und Gerätekombinationen für die terrestrische Aufnahme
Geräte und Gerätekombinationen für die photogrammetrische Auswertung und Interpretation (Komparatoren, Analogauswertegeräte, analytische Auswertegeräte, Stereoauswertegeräte, Differentialentzerrungsgeräte, Punktübertragungsgeräte)

528.711.114

Katushkov, V. A.
Primenenie svetoizluchajushhikh diodov dlja indikacii nomerov snimkov fototeodolitov
Geod. i Kartogr. — Moskva (1983) 6. — p. 45—47
The use of light emitting diodes for indication of phototheodolite photograph numbers (RUS)
AS 73.11
Key words: Phototheodolite

Dobrokhotov, National Correspondent, USSR IGD 84/III 175

528.711.11

Katushkov, V. A.
Razrabotka i issledovanie kamer dlja krupnomasshtabnojj s"jomki
Inzh. geod. – Kiev (1983) 26. – p. 22–25 : 2 lit.
Development and examination of cameras for large-scale surveys (RUS)
AS 73.11
Key words: Camera, large-scale survey

Dobrokhotov, National Correspondent, USSR IGD 84/III 176

www.ingramcontent.com/pod-product-compliance
Lightning Source LLC
LaVergne TN
LVHW091729021026
849161LV00003BB/503